AF413395

Order, Disorder and Criticality

Advanced Problems of Phase Transitions and Complex Systems

Volume 8

Order, Disorder and Criticality

Advanced Problems of Phase Transitions and Complex Systems

Volume 8

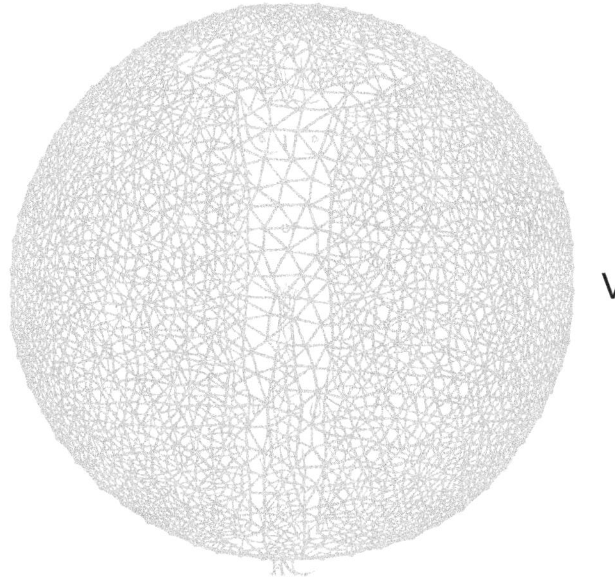

Editor

Yurij Holovatch

National Academy of Sciences, Ukraine

World Scientific

NEW JERSEY · LONDON · SINGAPORE · BEIJING · SHANGHAI · HONG KONG · TAIPEI · CHENNAI

Published by

World Scientific Publishing Co. Pte. Ltd.
5 Toh Tuck Link, Singapore 596224
USA office: 27 Warren Street, Suite 401-402, Hackensack, NJ 07601
UK office: 57 Shelton Street, Covent Garden, London WC2H 9HE

Library of Congress Control Number: 2024946027

British Library Cataloguing-in-Publication Data
A catalogue record for this book is available from the British Library.

Cover credit: The design of the cover incorporates the artwork of Jacques Hnizdovsky
(Constructor, Woodcut, 1967).

ORDER, DISORDER AND CRITICALITY
Advanced Problems of Phase Transitions and Complex Systems
Volume 8

For any available supplementary material, please visit
https://www.worldscientific.com/worldscibooks/10.1142/14039#t=suppl

ISBN 978-981-98-0081-0 (hardcover)
ISBN 978-981-98-0082-7 (ebook for institutions)
ISBN 978-981-98-0083-4 (ebook for individuals)

Preface

Twenty years ago, World Scientific Publishing started a new review volume series entitled *Order, disorder and criticality: Advanced problems of phase transition theory.*[1] The original goal of this series was to demonstrate that the theory of phase transitions and critical phenomena, whose golden age was in the last decades of the 20th century, did not recede into the past along with that century, but has continued to develop and that there is still much to be said there, both at the fundamental level and in terms of applications. This initiative turned out to be successful, and evidence of this, in particular, is the eighth volume of the series, which is being released this year.

Over time, the subject matter of the series has gone beyond discussing phase transitions and critical phenomena in condensed matter. The scope of the reviews have included analysis of tipping points and emergence phenomena in biological, social, technological systems of many interacting agents. Of particular interest were cases when the behavior of such systems as a whole did not follow trivially from the behavior of their components. In this way, the subject matter of the series (the contents of the previous volumes are given on pp. x–xiii) reflects the dynamics of the development of the complex system science. The latter increasingly uses the methods and concepts of physics, notably of statistical physics and the physics of phase transitions. Therefore, our review series has been designed not only to reflect this trend, but also to contribute to its success. Starting with this volume, we explicitly mention complex systems in the series subtitle, underscoring this fact.

As the previous volumes, the current one addresses traditional physical problems, as well as those where a physical approach is extended to the

---

[1] *Order, Disorder and Criticality: Advanced Problems of Phase Transition Theory,* edited by Yu. Holovatch (World Scientific, Singapore), vol. 1 – 2004, vol. 2 – 2007, vol. 3 – 2012, vol. 4 – 2015, vol. 5 – 2018, vol. 6 – 2020, vol. 7 – 2023.

beyond-physics field. It starts with the chapter *The enigmatic exponent ϙ and the story of finite-size scaling above the upper critical dimension* by Ralph Kenna and Bertrand Berche. Sadly, Ralph Kenna,[2] passed away last year, and the chapter has been finalized by Bertrand Berche, paying tribute to their long-standing collaboration and friendship. The chapter gives a rather personal story about the evolution of concepts of scaling, hyperscaling and finite-size scaling in high dimensions. The authors belong to those who shaped our modern understanding of these concepts; therefore, such a story is of special value. Their review describes the main steps taken to understand criticality above the upper critical dimension. The fundamental idea implemented by the finite-size scaling concept is that there are only two relevant length scales for a system of finite size: system length L and correlation length ξ. Near the critical point, it is assumed that the correlation length becomes equivalent to the entire size of the system. This, in turn, leads to the finite-size scaling relations. A bold and, at first glance, unusual statement was to relax the assumption that the correlation length has to be bounded by the physical length. It has been suggested that its growth with L is governed by the new critical exponent ϙ: $\xi \sim L^{ϙ}$. The growth is linear, below the upper critical dimension; however, it becomes superlinear at high dimensions, allowing to extend the validity of hyperscaling and finite-size scaling to the high-dimensional regime.

The perspective of phase transitions, scaling, and universality is also important for understanding the next chapter of this book – *Stochastic spatial Lotka–Volterra predator-prey models* by Uwe Täuber. This chapter discusses how methods of statistical physics in analytical and computational approaches allow an insight into and quantitative description of spontaneous pattern formation and other noise-induced phenomena occurring in the interacting populations. To this end, recent studies of the generalizations of the famous Lotka–Volterra predator-prey model[3] are considered. In the original setting, the dynamics of the predator-prey interplay has been described via a system of coupled mean-field rate equations. Yet another perspective opens, when one modifies the model to include stochasticity and spatial extension. This strategy has been successfully implemented by

---

[2] Ralph Kenna (August 27, 1964 – October 26, 2023) was an Irish scholar in the field of statistical physics and complex system science. See more about him in the special issues of the journals *Condensed Matter Physics* **27** (2024) and *Entropy* **26** (2024).

[3] As I write this preface in Lviv, Ukraine, I can't help but mention that Alfred J. Lotka was born here, and that the Ukrainian chemist Julian Hirniak, another Lviv native quoted by Lotka, was the proponent of periodic chemical reactions as early as in 1908. See more in N. Manz, Yu. Holovatch, J. Tyson, preprint arXiv:2407.19945 (2024).

the author of this chapter and his colleagues, having in mind applications in various fields ranging from physics and chemistry to biology, ecology, epidemiology, and even sociology. Both the models considered and the methods used in their analysis enable concentrating on the fluctuation and correlation effects that are not captured by the mean-field treatment. This, in turn, leads to subtle effects that can be interpreted as the far-from-equilibrium continuous phase transitions within the directed percolation universality class.

The third chapter of the book, *Statistical mechanics and artificial neural networks: Principles, models, and applications* by Lucas Böttcher and Gregory Wheeler, consists of two major parts. The first part can be read independently and serve as a short but self-consistent and self-contained introduction to the statistical mechanics of artificial neural networks (ANNs). As the authors themselves note, contributions to ANN research extend beyond statistical mechanics: "...Historically, these efforts have been marked by eras of interdisciplinary collaboration referred to as cybernetics, connectionism, artificial intelligence, and machine learning".[4] However, the role that statistical physics played and continues to play in these studies is difficult to overestimate, as the overview presented in this chapter aims to emphasize. Actually, some versions of artificial neural networks were inspired by the Ising model. This connection is explained in detail, taking the Hopfield model and Boltzmann machines as the examples and providing an overview of the other models, principles, and applications of ANNs. Such interpretation of concepts from the language of learning systems to that of statistical physics is of particular value for the uninitiated reader. The second part of this chapter discusses geometric properties of ANNs. Since the latter can be seen as high-dimensional mathematical functions, analysis of their landscapes in the high-dimensional space (the so-called loss landscapes) serves as an efficient tool to acquire information about their performance.

Whereas an approach to and a departure from a certain state is of equivalent relevance and is studied by using basically similar tools in settings of theoretical physics, this is not the case for a social system. As it is emphasized in the chapter *Political systems as complex, statistical systems – Unravelling the role of diversity and feedback for stability* by Karoline Wiesner, much research has focused on how democracies arise and can be facilitated; however, much less is known about the destabilization of

---

[4]See p. 118 of this book.

democracies. The latter phenomenon is termed in the political sciences as democratic backsliding. The chapter summarizes recent studies relating to these questions and shows how the complexity science and statistical mechanics are becoming useful tools in creating modern theories of political instability. Like the former chapter, this one also comprises two parts. The first part discusses inherent features of complex systems, relying on the taxonomy introduced in a recent book written in a cooperation with the current author.[5] Further discussion in the second part of the chapter focuses on three features of complex social systems, showing that their mutual relationship is of particular relevance for understanding the rise and fall of democracies. These are disorder, feedback, and stability. Qualitative conclusions of the discussion are supported by a recent quantitative study of data provided by the Varieties of Democracy project.

Concepts of scaling and universality are further discussed in the last chapter of the book, *Observing cities as a complex system* by Rafael Prieto-Curiel. Here, the quest for universality is exemplified by analyzing data about different cities. Specific constraints (history, geography, economy) are very important locally, but are there features of large cities that have some universal character? If so, what are these features and what do they indicate? The search for answers to these and similar questions has led to what is now called the science of cities, and the consideration of a city as a complex system is an important component of such research. In particular, the urban scaling theory gives a framework for analyzing cities in the context of their sizes. The chapter reviews some urban scaling principles as well as approaches for analyzing variations between different indicators for areas within the same city. In addition, the chapter provides an overview of the acceleration of the dynamics of global urbanization and provides numerous examples of studies of cities as complex systems.

This review series originates from the Ising Lectures,[6] an annual Workshop on critical phenomena and complex systems that started in Lviv in 1996. Year 2024 marks the centennial anniversary of Ernst Ising's doctoral thesis[7] defence and, with our book, we also mark this anniversary. I am

---

[5] J. Ladyman, K. Wiesner, *What Is a Complex System?* Yale University Press, New Haven & London, 2020.

[6] http://www.icmp.lviv.ua/ising/

[7] The now famous Ising model was described and solved in 1D in Ernst Ising's doctoral thesis written under supervisorship of Wilhelm Lenz: *Beitrag zur Theorie des Ferro- und Paramagnetismus. Dissertation zur Erlangung der Doktorwürde der Mathematisch–Naturwissenschaftlichen Fakultät der Hamburgischen Universität vorgelegt von Ernst Ising aus Bochum. Hamburg 1924.*

deeply thankful to the authors of this volume for lecturing at the Workshop and for submitting their contributions for this book. Special thanks are due to Lakshmi Narayanan from World Scientific, Singapore — her advice and friendly attitude have accompanied me from the first volume of this series — and to Yulian Honchar from the ICMP, Lviv for his help in the technical compilation of this book.

Yurij Holovatch
Institute for Condensed Matter Physics,
National Academy of Sciences of Ukraine, Lviv
and
$\mathbb{L}^4$ Collaboration & International Doctoral College
for the Statistical Physics of Complex Systems,
Leipzig-Lorraine-Lviv-Coventry, Europe

28.06.2024

Contents of vol. 1

Mathematical Theory of the Ising Model and its Generalizations: an Introduction
Yuri Kozitsky

Relaxation in Quantum Spin Chains: Free Fermionic Models
Dragi Karevski

Quantum Phase Transitions in Alternating Transverse Ising Chains
Oleg Derzhko

Phase Transitions in Two-Dimensional Random Potts Models
Bertrand Berche and Christophe Chatelain

Scaling of Miktoarm Star Polymers
Christian von Ferber

Field Theoretic Approaches to the Superconducting Phase Transition
Flavio S. Nogueira and Hagen Kleinert

Contents of vol. 2

Introduction to the Non-Perturbative Renormalization Group
Bertrand Delamotte

Introduction to Critical Dynamics
Reinhard Folk

Spacetime Approach to Phase Transitions
Wolfhard Janke and Adriaan M. J. Schakel

Representations of Self-Organized Criticality
Alexander Olemskoi and Dmytro Kharchenko

Phase Transitions in the Pseudospin-Electron Model
Ihor Stasyuk

Contents of vol. 3

Universal Scaling Relations for Logarithmic-Correction Exponents
Ralph Kenna

Phase Behaviour and Criticality in Primitive Models of Ionic Fluids
Oksana Patsahan and Ihor Mryglod

Monte Carlo Simulations in Statistical Physics – From Basic Principles to Advanced Applications
Wolfhard Janke

Ising Model on Connected Complex Networks
Krzysztof Suchecki and Janusz A. Hołyst

Minority Game: An "Ising Model" of Econophysics
František Slanina

Contents of vol. 4

Scaling and Finite-Size Scaling above the Upper Critical Dimension
Ralph Kenna and Bertrand Berche

Monte Carlo Simulations of Critical Casimir Forces
Oleg A. Vasilyev

Non-ergodicity and Ageing in Anomalous Diffusion
Ralf Metzler

Kinetics of Pattern Formation: Mesoscopic and Atomistic Modelling
Helena Zapolsky

A Renormalization Group Like Model for a Democratic Dictatorship
Serge Galam

Contents of vol. 5

Statistical Properties of One-Dimensional Directed Polymers in a Random Potential
Victor Dotsenko

Non-Euclidean Geometry in Nature
Sergei Nechaev

Dynamics of Polymers: Classic Results and Recent Developments
Mikhail V. Tamm and Kiril Polovnikov

Generalized Ensemble Computer Simulations of Macromolecules
Wolfhard Janke

Photo-Controllable Networks in Macromolecular Solutions and Blends
Jaroslav M. Ilnytskyi

Monte Carlo Methods for Massively Parallel Computers
Martin Weigel

Complex Networks and Infrastructural Grids
Antonio Scala

Contents of vol. 6

Nature of the Spin Glass Phase in Finite Dimensional (Ising) Spin Glasses
Juan J. Ruiz-Lorenzo

Phase Transitions in Fermi Systems
Paweł Jakubczyk

Geometrical Frustration in ISAW Models
Damien P. Foster

Two-Dimensional Systems of Elongated Particles
Nikolai I. Lebovka and Yuri Yu. Tarasevich

A Simple Model of Tumor Growth
Yuri Kozitsky and Krzysztof Pilorz

Models for Vehicular and Pedestrian Flows
Massimiliano D. Rosini

Contents of vol. 7

The Survival of Ernst Ising and the Struggle to Solve His Model
Reinhard Folk

Insights into the Collapse of Adaptive Complex Systems
Leonhard Horstmeyer

Some Old and New Puzzles in the Dynamics of Fluids
Ihor Mryglod and Vasyl Ignatyuk

Phase Transitions and Criticality in the Collective Behavior of Animals – Self-Organization and Biological Function
Pawel Romanczuk and Bryan C. Daniels

Approaches to the Classification of Complex Systems: Words, Texts, and More
Andrij Rovenchak

Contents

Preface   v

1. The Enigmatic Exponent ϟ and the Story of Finite-Size
    Scaling Above the Upper Critical Dimension   1

    R. Kenna and B. Berche

2. Stochastic Spatial Lotka–Volterra Predator-Prey Models   67

    U. C. Täuber

3. Statistical Mechanics and Artificial Neural Networks:
    Principles, Models, and Applications   117

    L. Böttcher and G. Wheeler

4. Political Systems as Complex, Statistical Systems – Un-
    ravelling the Role of Diversity and Feedback for Stability   163

    K. Wiesner

5. Observing Cities as a Complex System   181

    R. Prieto-Curiel

Index   213

Chapter 1

The Enigmatic Exponent $\digamma$ and the Story of Finite-Size Scaling Above the Upper Critical Dimension

Ralph Kenna*

*Centre for Fluid and Complex Systems,
Coventry University, Coventry, CV1 5FB, United Kingdom
$\mathbb{L}^4$ Collaboration & Doctoral College for the Statistical Physics of Complex
Systems, Leipzig-Lorraine-Lviv-Coventry, Europe*

Bertrand Berche[†]

*Laboratoire de Physique et Chimie Théoriques,
Université de Lorraine - CNRS, UMR 7019, Nancy, B.P. 70239, F-54506
Vandœuvre les Nancy, France
$\mathbb{L}^4$ Collaboration & Doctoral College for the Statistical Physics of Complex
Systems, Leipzig-Lorraine-Lviv-Coventry, Europe*

Scaling, hyperscaling and finite-size scaling were long considered problematic in theories of critical phenomena in high dimensions. The scaling relations themselves form a model-independent structure that any model-specific theory must adhere to, and they are accounted for by the simple principle of homogeneity. Finite-size scaling is similarly founded on the fundamental idea that only two length scales enter the game — namely system length and correlation length. While all scaling relations are quite satisfactory for multitudes of physical systems in low dimensions, one fails in high dimensions. The aberrant scaling relation is called hyperscaling and involves dimensionality itself. Finite-size scaling also appears to fail in high dimensions. Developed in the 1930s, Landau mean-field theory is valid in such high-dimensional systems. However, it too does not accord with hyperscaling and finite-size scaling there. The advent of renormalization-group theory in the 1970s brought deeper fundamental insights into critical phenomena, allowing systems to be viewed at different scales. Above a critical dimensionality, higher-order Renormalization Group (RG) eigenvalues become irrelevant and scaling is governed by the Gaussian fixed point. Although obeying all scaling relations including hyperscaling, and although it appears to successfully explain

*ORCID: https://orcid.org/0000-0001-9990-4277
[†]bertrand.berche@univ-lorraine.fr, ORCID: https://orcid.org/0000-0002-4254-807X

scaling in the correlation sector, the Gaussian fixed point fails to capture the free energy and derivatives, even in infinite volume. In the 1980s, to fix this for the magnetisation, specific heat and susceptibility, Fisher introduced the notion of dangerous irrelevant variables. Since the correlation sector did not appear to be broken, no attempt was made to repair it and Fisher's modified RG formalism worked quite well in the thermodynamic limit of infinite volume. However, finite-size scaling fails. Also in the 1980's, Binder, Nauenberg, Privman and Young extended Fisher's concept to the free energy itself and to finite-size systems. While putting Fisher's ideas on a more fundamental footing, the failure of finite-size scaling there still presented a problem. This appeared to be resolved by the introduction of "thermodynamic length" to replace correlation length as the length scale that controls Finite-Size Scaling (FSS). Thus hyperscaling and FSS were both sacrificed in favour of the RG patched together by ad hoc solutions. In the 1990's, Luijten and Blöte went a long way to resolving the dilemma by adding corrections to scaling to the above considerations. It was clear that both of these played a role. However, as with previous authors, and adhering to the principle of not fixing that appears not to be broken, they did not address correlation length directly.

Here we report on developments over the past decade which went a long way to addressing these long-standing problems The key to unlocking these, and extending their validity to the high-dimensional regime, was to relax assumptions that the correlation length has to be bounded by the physical length of bounded systems. This allowed and necessitated the extension of Fisher's concept of dangerous irrelevant variables to the correlation sector.

Contents

1. Introduction 4
2. Homogeneity, scaling, logarithms and the enigmatic ϙ 10
    2.1. Homogeneity 10
    2.2. Logarithmic corrections 13
    2.3. The enigmatic nature of $\hat{\qoppa}$ and of $\qoppa$ 17
    2.4. Summary 18
3. The fundamental theory of phase transition as a framework 19
    3.1. Lee-Yang zeros 21
    3.2. Fisher zeros 25
    3.3. Pseudocriticality, shifting and rounding 27
    3.4. Summary 29
4. Ginzburg-Landau theory (mean-field) 31
5. The Gaussian fixed point: its apparent sufficiency for the correlation sector and its insufficiency for the free-energy sector 36
6. Dangerous irrelevancy: Rescuing the free-energy sector in infinite volume and attempts at resuscitation in finite volume 38

6.1.   Dangerous irrelevant variables for thermodynamic functions at infinite
volume 38
6.2.   A problem with Finite-Size Scaling 40
6.3.   Dangerous Irrelevant Variables for finite size: the "starred" Renormalization Group eigenvalues 41
7.   Scaling of the Fourier modes 43
8.   Corrections to finite-size scaling and their dominance at short distances 44
9.   Extension of dangerous irrelevancy in the correlation sector: the use of ϙ, the
rescue of hyperscaling and finite-size scaling 50
10. The case of Free Boundary Conditions 53
10.1. A mismatch between T_L and T_c in free boundary conditions 53
10.2. The three possible scenarios 55
10.3. Towards a Gaussian Fixed Point scaling at T_c 56
11. Conclusion 57
References 59

To the memory of my friend, Ralph Kenna

Ralph and I started writing this review article for a volume of the series edited by our common friend Yurij Holovatch while Ralph was already ill. He knew that he would not see the book published. We had worked a bit earlier on a lecture format published in Scipost,[1] and this was not an easy task to propose a different perspective. This is Ralph's idea to write a kind of story, with many personal views and quotations. The material presented here covers about fifteen years of our collaboration, and there is a lot to say. The result is maybe not really a scientific paper in the usual sense. As I wrote above this is more like a story, as the title says already. Unfortunately, Ralph passed away before this was finished. The reader will easily recognize which parts were written by Ralph (the good ones), and which are mine (those with little originality).

Those who met Ralph knew that he was a wonderful person. He was always enthusiastic, he was bringing new ideas onto the scene and was ready to fight, in a positive sense, in support of his opinions. He has been an excellent friend and an outstanding collaborator. We miss him.

This article is for a volume edited by our Ukrainian friend, (YuH). At the time I am finishing it, after more than two years of war imposed on his country by Putin's government of Russia, another terrible war, imposed by Netanyahu's government of Israel, is devastating the Gaza strip, denying the right of Palestinian people to live there. Ralph is no longer there to tell me how he would feel bad to witness such a horrible situation in the Middle East, but I know how he, like many in Ireland, was against any kind of apartheid and colonisation. This is a hard time for the world when international laws and international regulations of Human Rights are trampled in the complete disinterest of most countries. Many of these countries claim at the same time that they are democracies and claim that they work for Human Rights. Those countries that sell weapons to destroy Gaza. As simple human beings, we are against any kind of discrimination, any kind of racism, antisemitism, or islamophobia. Today my thoughts are in support of the Palestinian people who are suffering a tragedy imposed by Netanyahu's government of Israel. History and the international courts will qualify these mass killings as they should be. Our thoughts have been on Israeli victims of the hideous attacks of the 7th of October, today, they are for Palestinian children.[a]

April 2024

1. Introduction

In the proceedings of a conference held in Washington DC in April 1965, Michael Fisher published what is an important but widely forgotten paper titled "Notes, definitions, and formulas for critical point singularities".[2] Prior to Fisher's intervention, different people used different symbols at different times for different measures. Fisher pointed to us this when he invited us to meet him at the Royal Society in London, just before his retirement in October 2012. The paper opens with the introduction:

> "For the convenience of participants a set of definitions and notes concerning the various critical point singularities and a table of formulas relating the lattice gas and Ising ferromagnetic models were distributed at the Conference. These are reproduced here with a few corrections and extensions."

This is how the worldwide practice of using α, β, γ, δ, η and ν (arranged in that order in Fisher's paper) was introduced and accepted in the community for critical exponents of the specific heat, spontaneous magnetisation, susceptibility, induced magnetisation, anomalous dimension and correlation length. Of course, phase transitions can only happen for infinite-sized systems and there is no mention of finite volume in Fisher's paper; the term "critical point singularities" in the title makes that clear.

---

[a] https://www.amnesty.org/en/latest/news/2024/02/israel-opt-new-evidence-of-unlawful-israeli-attacks-in-gaza-causing-mass-civilian-casualties-amid-real-risk-of-genocide/

In his paper, Fisher defined a generic critical exponent λ in the following way:

> "if
> $$\lim_{x \to 0^+} \log f(x)/\log x = \lambda$$
>
> we may say
> $$f(x) \sim x^\lambda \text{ as } x \to 0^+.$$
>
> Note that with this definition the statement $f(x) \sim x^\lambda$ does not exclude the possibility of a logarithmically divergent factor, *i.e.*, $Ax^\lambda |\log x| \sim x^\lambda$. In particular if $\lambda = 0$ the function $f(x)$ might diverge as $|\log x|$, for example, or $f(0)$ might be finite, the function $f(x)$ then being either continuous or discontinuous as x passes through zero."

This is an example of the meticulous attention to detail characteristic of Fisher; he had not forgotten logarithmic corrections as an essential feature in the physics of critical phenomena. Although explicitly mentioning logarithmic corrections to the specific heat when α is zero in the two-dimensional Ising model, for example, he did not address them further in his paper. He did, however, distinguish between critical exponents at either side of the phase transition, using primed exponents for the broken phase and unprimed ones for the symmetric phase. Thus α was used as the critical exponent for the specific heat when temperature T exceeds its critical value T_c, for example, while α' was used for $T < T_c$.

This is why when one of us reported on the exponents of logarithmic corrections in Volume III of this series (2012), hatted, rather than primed notations, were used.[3] Thus if $f(x)$ has a logarithmic correction it is expressed as $f(x) \sim x^\lambda |\ln x|^{\hat{\lambda}}$ with $\hat{\alpha}$, $\hat{\beta}$, $\hat{\gamma}$, $\hat{\delta}$, $\hat{\eta}$ and $\hat{\nu}$ representing exponents of logarithmic corrections for the observables listed above. Just as there are scaling relations between the leading critical exponents, so too are there relations between their logarithmic counterparts. Indeed, it was the introduction of these scaling relations in Refs. [4,5] (and in Volume III of this series) that captured Fisher's attention and led to our memorable invitation to meet him at the Royal Society.

However, the analogue is not perfect; for the scaling relations between the hatted exponents to match exact and numerical calculations for specific models, there needed to be a logarithmic correction to the finite-size behaviour of the correlation length. This was termed $\hat{q}$ in Refs. [3–5] and had no counterpart in Fisher's list. Indeed, it referred to finite systems where a singularity cannot emerge, so that if L denotes the linear extent of a finite system, the correlation length scales as $\xi_L \sim L(\ln L)^{\hat{q}}$. Serendipitously, the missing counterpart emerged the same year as our meeting with Fisher in our paper about hyperscaling above the upper critical dimension published as Ref. [6]. Since we had used $\hat{q}$ for the exponent governing logarithmic corrections to finite-size scaling [FSS] of the correlation length, it was natural to use q as its leading counterpart so that $\xi_L \sim L^{q}$.[b]

"There is one thing wrong with your theory" were the terrifying words of Michael Fisher at our meeting — "the letter 'q' is not right". It did not match the Greek alphabet! Our relief that the deficiency in our theory rested only with nomenclature, and not in bringing finite-size concepts into descriptions of (infinite-volume) critical-point singularities, emboldened us to offer an excuse that all the Greek letters had been used up! This was met with calm assurance that again revealed Fisher's meticulousness as he etched out the symbol "ϙ": "Use 'coppa'," he said. "It's an archaic Greek letter." Thus a new symbol entered the pantheon tamed by Fisher nearly half a century earlier.

Our claims of a superlinear correlation length in high dimensions raised eyebrows in some quarters, as did the usage of ϙ. Resistance to the former was due to dogmatic belief that the correlation length dare not exceed the length of a finite system, despite this already having been proved in specific models twenty years earlier by Édouard Brézin.[7] Hostility to the nomenclature is captured in the statement: "The authors' introduction of their new exponent with the ankh notation is thus a gimmick which is not recommended in the scientific literature. Instead, they should perhaps find a new letter that can be easily typeset." Faced with the options of defying that anonymous referee or Michael Fisher, we did not waver and *A new critical exponent ϙ and its logarithmic counterpart ϙ̂* was published in Condensed Matter Physics,[c] John Cardy having come to the rescue for the typesetting.[8]

---

[b]The symbol $\hat{q}$ was used in Ref. [4] simply because q is next in the alphabet after p, which was used instead of the gap exponent for scaling of Lee-Yang zeros in an earlier version of that paper. This usage of p was intended to indicate that the theory set out in Ref. [4] was not reliant on concepts such as renormalization-group eigenvalues. While the eventual usage of Δ instead of p preempts the double meaning of "gap" (both meanings are explained below), the consequent usage of q and $\hat{q}$ was serendipitous, as it fits very well with the convolution of related literature, as we shall see below.

[c]The journal was founded in 1993 by the Institute for Condensed Matter Physics of the National Academy of Sciences of Ukraine. In March 2022 they issued the statement "Our country has suffered an unjust and unprovoked aggression. Our army and our people are carrying an existential fight for Ukraine's very survival as an independent state, for our lives and freedom. Despite everything, we here, not at the frontline, are determined to do our job and continue publishing new issues of our journal on a regular basis, as long as it is possible. If you wish to support the Ukrainian scientific community, we are cordially inviting you to submit your manuscripts to our journal." We refer the reader to the journal's website: http://www.icmp.lviv.ua/journal.

Hyperscaling, which is one of the four scaling relations that interlink the six most prominent critical exponents α, β, γ, δ, ν and η, usually reads as

$$\nu d = 2 - \alpha. \tag{1}$$

Here, d is the dimensionality of the system and does not feature in the other three scaling relations (which we list in Subsection 2.1, below). The Ginzburg criterion (explained below) marks the critical value d_{uc} of dimensionality beyond which critical fluctuations of the order parameter cease to play a leading role. Mean-field theory (MFT) is a valid first approximation there and the criterion dictates that ν and γ are fixed to their critical-dimension values. The remaining critical exponents, α included, are fixed by the remaining scaling relations. With α and ν so fixed, and d free to roam above its critical value d_{uc} Eq.(1) cannot hold there and, for this reason, it is frequently said (including in textbooks, reviews and even new literature) that FSS fails above the upper critical dimension.

The introduction[8] of the new exponent $\digamma$ renders such statements obsolete; a small alteration to the expressions (1) based on a foundational alteration to the renormalization group, delivers a new form which is valid above (and below) d_{uc}. In volume IV of this book series, we provided a renormalization-group basis for the correlation length ξ_L to take the form

$$\xi_L \sim L^{\digamma}, \tag{2}$$

with

$$\digamma = \frac{d}{d_{\mathrm{uc}}} \tag{3}$$

above d_{uc} and $\digamma = 1$ below. The new hyperscaling relation is then

$$\nu \frac{d}{\digamma} = 2 - \alpha. \tag{4}$$

This statement of hyperscaling is far from trivial - it is no mere "gimmick" or cosmetic alteration of Eq.(1) to phenomenologically render $\nu d_{\mathrm{uc}} = 2 - \alpha$. It is rooted in a very subtle and fundamental extension of the renormalization group to the correlation sector. The subtlety of this extension is evinced by it having lain hidden for nearly fifty years — precisely because of the success and simplicity of mean-field theory.

As stated, the superlinear behaviour of the correlation length was already introduced in Volume III of this series in a renormalization-group-independent manner.[3] Ref. [3] reported on a series of papers,[9–13] where one of us presented a self-consistent theory of finite-size scaling at the upper critical dimension itself. There, logarithmic corrections modify mean-field behaviour in an essential way. Expressed into $\hat{\digamma}$ instead of into $\hat{q}$, the counterparts of Eq.(2) and Eq.(3) are

$$\xi_L \sim L(\ln L)^{\hat{\digamma}},\tag{5}$$

with

$$\hat{\digamma} = \frac{1}{d_{\mathrm{uc}}}.\tag{6}$$

These led to a logarithmic counterpart of the new hyperscaling relation (4), namely

$$\hat{\nu} = \hat{\digamma} - \frac{\hat{\alpha}}{d_{\mathrm{uc}}}.\tag{7}$$

One of the objectives of this chapter is to report on the origins of and necessity for the above formulae for hyperscaling — over fifty years after the old version was first set down.[14–17] Eq.(6) holds in most cases but there are exceptions — see Refs. [18,19] for details. (We do not require those details for what is to come.)

Finite-size scaling is also considered to fail above the upper critical dimension d_{uc}. Its standard form, valid below (but not above) the upper critical dimension, reads

$$\frac{Q_L(\tau = 0)}{Q_\infty(\tau)} = \mathcal{F}_Q\left[\frac{L}{\xi_\infty(\tau)}\right].\tag{8}$$

Here Q is a generic thermodynamic function such as specific heat c or magnetic susceptibility χ. The variable

$$\tau = T/T_c - 1\tag{9}$$

represents the reduced temperature of the system and vanishes at the critical point of its infinite-volume thermodynamic limit. The subscripts in Eq.(8) represent the system size and ξ is the correlation length. A hand-waving argument often used in favour of the FSS form (8) is that L and ξ are the only two length scales associated with a given system, so all forms of scaling should depend on their ratio. If $Q_\infty(\tau)$ diverges as the critical point is approached as $|\tau|^{-\rho}$, say, this divergence has to be matched by the functional dependence of $\mathcal{F}_Q$ on the correlation length. This delivers the classic FSS form

$$Q_L(0) \sim L^{\rho/\nu},\tag{10}$$

which is not divergent for finite L.

In the case of magnetic susceptibility, this gives $\chi_L(0) \sim L^{\gamma/\nu}$. However, the FSS ratio γ/ν is not matched by Brézin's exact calculations and numerical simulations by Kurt Binder for specific models above the upper critical dimension.[7,20] In particular, already in the 1980's both methods showed that $\xi_L \sim L^{5/2}$ for the 5D Ising model, rather than L^2 which is what standard FSS with $\gamma = 1$ and $\nu = 1/2$ would predict. Eq.(10) therefore fails above the upper critical dimension and this is frequently stated in the same textbooks, reviews and other literature that condemn hyperscaling to failure there. A primary outcome of our chapters in Volumes III and IV of this series is the replacement of Eq.(8) by

$$\frac{Q_L(t=0)}{Q_\infty(t)} = \mathcal{F}_Q\left[\frac{\xi_L(t=0)}{\xi_\infty(t)}\right], \tag{11}$$

where

$$t = \frac{T}{T_L} - 1, \tag{12}$$

in which T_L is a pseudocritical point. We refer to this as QFSS to distinguish it from standard FSS. There are two important changes in QFSS of Eq.(11) relative to standard FSS in Eq.(8). The first is the replacement ξ_L/ξ_∞ for L/ξ_∞. The necessity for this change was already suggested in 1991 when, in Ref. [9], it was used to theoretically identify and numerically test FSS for Fisher zeros and the specific heat for the ϕ^4 model in four dimensions. The second change is the replacement[6] of reduced temperature $\tau = T/T_c - 1$ by $t = T/T_L - 1$ (see also Volume IV).[21]

As stated, in Volume III, narrowly preceding our 2012 Royal-Society meeting, $\hat{q}$ was used rather than Fisher's $\hat{ϟ}$. Three years later, in Volume IV (2015), we presented a chapter on scaling above the upper critical dimension centered on ϕ^4 theory and Ising models with short- and long-range interactions. There the new archaic symbol $ϟ$ (and $\hat{ϟ}$) was deployed. In the meantime, the notation has been taken up in multiple studies of phase transitions in high dimensions, *e.g.* recently in Refs. [22–25].

Here, in Volume VIII, we again deploy the notation suggested by Fisher. The purpose of this chapter is to generalise and extend some of the concepts presented previously. We start by introducing very fundamental (model-independent) Widom scaling concepts and extending them to the correlation sector when logarithmic corrections are present. This latter move will be justified by posteriori, by a similar extension to the renormalization-group formalism. These extensions are a main contribution of this body of work.[1,3–6,8–13,18,21,26–28] On this basis, we re-derive three of the scaling relations for logarithmic corrections. We also introduce the fundamental theory of phase transitions, finite-size scaling, and pseudocritical concepts. This provides a base for us to build upon Volumes III and extend Volume IV to the more general case of ϕ^n theory.

2. Homogeneity, scaling, logarithms and the enigmatic ϙ

> "It should be clear that the purely thermodynamic approach employed in this paper cannot in and of itself tell us the behaviour of various functions near the critical point. Rather, it provides correlations among data obtained from experiments, or from statistical calculations, and checks on their consistency. The value of thermodynamics for checking consistency has already been demonstrated."

These are the words of Robert Griffiths whose name, through his paper,[16] is associated with one of the scaling laws for critical exponents. The same can be said for Refs. [4,5] on which Section 3 of this chapter rests. The scaling theory presented there is based on self-consistencies, which are manifest as relations between the various exponents associated with logarithmic correction to scaling and finite-size scaling. For ab initio model-specific theories, the renormalization group and related approaches are appropriate. That was the subject of our contribution to Volume IV of this series. There we addressed the ϕ^4 model, a model that trims back to the bare essentials of dimensionality and symmetry — the essence of universality.

In the usual spirit of the renormalization group, we consider a model with even and odd control parameters or fields τ and h. The even field τ may be thought of as a reduced temperature in the case of spin models: $\tau = T/T_c - 1$. The odd ordering field h can be thought of as $h = \beta H$ where $\beta = 1/k_B T$ is the inverse temperature up to the Boltzmann constant k_B. In Ising-type models, H is the external magnetic field. In the case of percolation theory, for example, where there is no temperature or magnetic field the probability of site occupation plays the role of β.

2.1. *Homogeneity*

A homogeneous function $f(\tau, h)$ is one that is only affected by a multiplicative rescaling if τ and h themselves are multiplicatively rescaled. The Widom scaling hypothesis is that the singular part of the free energy density of ϕ^n theory is so governed and

$$f_\infty(\tau, h) = b^{-d} f_\infty(b^{y_t}\tau, b^{y_h}h), \tag{13}$$

where b is an arbitrary rescaling factor.[29,30] The precision that this is the singular part of the free energy density is important as we know that regular contributions do also exist but they do not enter the following discussion.

Here the subscript indicates we are dealing with a system of infinite extent in all d directions. A similar hypothesis may be applied to the correlation function and the correlation length so that

$$g_\infty(\mathbf{x}, \tau, h) = b^{-2x_\phi} g_\infty(b^{-1}\mathbf{x}, b^{y_t}\tau, b^{y_h} h) \tag{14}$$

$$\xi_\infty(\tau, h) = b\,\xi_\infty(b^{y_t}\tau, b^{y_h} h). \tag{15}$$

Here $\mathbf{x}$ is a measure of physical distance in any direction of the d-dimensional system and isotropy is assumed.

If the theory exhibits a phase transition at $T = T_c$ or $\tau = 0$ and $H = h = 0$ between a low-temperature phase which is ordered and a high-temperature one which is disordered, the two phases are distinguished by the value of an order parameter, and, at this critical point, physically measurable properties become singular. These thermodynamic quantities (magnetization, internal energy density, magnetic susceptibility and specific heat) are derivable from the free energy density as follows:

$$m_\infty(\tau, h) = \frac{\partial f_\infty(\tau, h)}{\partial h} = b^{-d+y_h} m_\infty(b^{y_t}\tau, b^{y_h} h), \tag{16}$$

$$e_\infty(\tau, h) = \frac{\partial f_\infty(\tau, h)}{\partial \tau} = b^{-d+y_t} e_\infty(b^{y_t}\tau, b^{y_h} h), \tag{17}$$

$$\chi_\infty(\tau, h) = \frac{\partial^2 f_\infty(\tau, h)}{\partial h^2} = b^{-d+2y_h} \chi_\infty(b^{y_t}\tau, b^{y_h} h), \tag{18}$$

$$c_\infty(\tau, h) = \frac{\partial^2 f_\infty(\tau, h)}{\partial \tau^2} = b^{-d+2y_t} c_\infty(b^{y_t}\tau, b^{y_h} h). \tag{19}$$

In these relations, signs and multiplicative factors that do not compromise the form of the singularities are omitted.

Thus the simple homogeneity assumption allows the various thermodynamic functions to be expressed in terms of only two scaling dimensions y_t and y_h. We may express them in the conventional form that predates the renormalization group as follows.[2] Firstly, setting $\tau = 0$ and $b = |h|^{-1/y_h}$,

$$m_\infty(0, h) \simeq D_c^{-1/\delta}|h|^{1/\delta}, \quad \delta = \frac{y_h}{d - y_h}, \quad D_c^{-1/\delta} = m_\infty(0, 1). \tag{20}$$

Then, in zero magnetic field, setting $b = |\tau|^{-1/y_t}$ delivers[d]

$$m_\infty(\tau,0) \simeq B^-|\tau|^\beta, \ \beta = \frac{d - y_h}{y_t}, \quad B^- = m_\infty(-1,0), \ \tau < 0, \quad (21)$$

$$\chi_\infty(\tau,0) \simeq \Gamma^\pm|\tau|^{-\gamma}, \ \gamma = \frac{2y_h - d}{y_t}, \quad \Gamma^\pm = \chi_\infty(\pm 1,0), \quad (22)$$

$$c_\infty(\tau,0) \simeq \frac{A^\pm}{\alpha}|\tau|^{-\alpha}, \ \alpha = \frac{2y_t - d}{y_t}, \quad \frac{A^\pm}{\alpha} = c_\infty(\pm 1,0). \quad (23)$$

Here, we omit non-universal metric factors (see Refs. [31–33]). We can apply a similar convention to the correlation function and correlation length to obtain

$$g_\infty(\tau = 0, h = 0, \mathbf{x}) \sim \frac{1}{|\mathbf{x}|^{d-2+\eta}}, \quad (24)$$

and

$$\xi_\infty(\tau,0) \simeq \Xi^\pm|\tau|^{-\nu}, \ \nu = \frac{1}{y_t}, \quad \Xi^\pm = \xi_\infty(\pm 1,0), \quad (25)$$

$$\xi_\infty(0,h) \simeq \tilde{\Xi}^\pm|h|^{-\nu_c}, \ \nu_c = \frac{1}{y_h}, \quad \tilde{\Xi}^\pm = \xi_\infty(0,\pm 1). \quad (26)$$

Eliminating the scaling dimensions y_t and y_h in favour of the measurable critical exponent delivers the scaling relations

$$\nu d = 2 - \alpha, \quad (27)$$

$$\alpha + 2\beta + \gamma = 2, \quad (28)$$

$$\beta(\delta - 1) = \gamma, \quad (29)$$

$$\nu(2 - \eta) = \gamma. \quad (30)$$

These famous scaling relations were developed in the 1960's through altogether different methods by Benjamin Widom,[14–17] Leo Kadanoff,[30] John Essam and Michael Fisher.[34,35] The first three are most commonly referred to as Josephson's,[36] Rushbrooke's[37] and Griffiths'[38] scaling laws, in honour of their eponymous proponents, and the fourth was developed by Fisher.[39,40] The first and last scaling relations are conspicuous in that they involve the dimension of the system d and the anomalous dimension η. The first formula here recovers Eq.(1) and, although named *hyperscaling*, it does not hold for a hypercube such as the Ising model above four dimensions. As stated earlier, it only holds at and below the upper critical dimension. The

---

[d]Repeated differentiation wrt field delivers higher moments so that the nth magnetic moment scales as $|\tau|^{(d-ny_h)/y_t}$. The "gap" between exponents is then $y_h/y_t = \beta + \gamma = \beta\delta$ and is conventionally given the label Δ.

quantity η in the last scaling relation is called the *anomalous dimension* because its deviation from zero is a measure of deviation from mean-field theory. It governs the decay with distance of the correlation function as

$$d - 2 + \eta = 2x_\phi. \tag{31}$$

David Nelson, in his remembrances,[41] reports:

> "Fisher is uniquely responsible for the prediction of the anomalous critical exponent η (Michael once owned a boat that sailed on Lake Cayuga in Ithaca, New York, named the Eta), which controls the decay of order parameter correlations at the critical temperature. This exponent also plays a key role in quantum field theories, where it is closely related to the anomalous scaling dimensions. As Michael himself once said, the exponent is numerically small, but nevertheless quite important."

We also find

$$\nu_c = \frac{\delta - 1}{2\delta}. \tag{32}$$

The last formula is less commonly used and does not have a name (that we know of). We display it distinctly here as it hides a subtlety emblematic of some of the features we wish to bring to the fore in this review.

To summarise, a simple application of the homogeneity assumption (before the Renormalization Group (RG) comes into play) for the free energy, correlation function and correlation length, enables the critical exponents α, β, γ, δ, ν, η, as well as ν_c to be written in terms of scaling dimensions y_t, y_h and x_ϕ as well as the dimensionality d. This enables us to write down the scaling relations which were derived by other means in the 1960's (before the advent of the renormalization group). We have not yet discussed the actual values of the scaling dimensions or critical exponents as these are model-specific. We repeat that exact and numerical results show that hyperscaling in the form (27) fails above the upper critical dimension. Next, we do similar for logarithmic corrections — we refer to the literature for their independent derivation and re-derive them from the homogeneity assumption.

2.2. *Logarithmic corrections*

At the critical dimension $d = d_{\mathrm{uc}}$, logarithmic corrections arise. These were presented using a similar formalism to that used above for ϕ^4 theory over 20 years ago by Nevzat Aktekin.[42] The introduction of logarithms, of course, loses the homogeneous properties of the free energy and related functions

— multiplicative rescaling of the argument of a logarithm does not result in multiplicative rescaling of the logarithm itself! Notwithstanding this, inspired by the Privman-Fisher form (13) Aktekin proposed the following formula for the 4D Ising model ($n = 4$):

$$f_L^{\rm s}(\tau, h) = L^{-4} Y[L^2 (\ln L)^{1/6} \tau, L^3 (\ln L)^{1/4} h].\tag{33}$$

with $\tau = T/T_c - 1$. From this, FSS for the thermodynamic functions ensue, finding support in Ref. [42] by Monte Carlo simulations on simple four-dimensional lattices linear extent up to $L = 16$ with periodic boundary conditions. Aktekin justifies this structure by referring to previous work by Erik Luijten and Henk Blöte:[43] "for the Ising model in $d = d_{\rm uc}$, the expression for $f_L^{\rm s}(\tau, h)$ derived starting with the renormalization group equations in differential form[43] reduces to the one given in Eq.(33) for $L \to \infty$ and confirms it." We return to Luijten and Blöte's seminal contribution in Section 8 below.

The contributions of homogeneous and inhomogeneous terms to the scaling formalisms of the renormalized Schwinger functions for the ϕ^4 theory (including finite size) were also described in Refs. [10,44,45]. Below four dimensions, the inhomogeneous term is not divergent in the thermodynamic limit. There the homogeneous term leads to critical behaviour (as captured in Subsection 2.1 above). As we shall see, the Gaussian fixed point is unstable for $d < 4$ and does not govern large distance behaviour. Instead, the Wilson-Fisher fixed point governs critical behaviour. In contrast, the Gaussian fixed point is stable above the critical dimension four. As the critical dimension is approached, merged fixed point at the origin leads to a double zero of the so-called Callan–Symanzik beta function and this is responsible for the occurrence of logarithmic corrections. The inhomogeneous term then contributes to the leading singular behaviour too. The homogeneous term remains singular, however, and, as shown in Ref. [10], contributes to divergences such as that in the susceptibility. Thus, while movement of the Wilson-Fisher fixed point to its Gaussian counterpart is responsible for triviality, the double zero gives rise to logarithmic corrections.

Keeping focus on the inhomogeneous term (but not forgetting the well-known role still played by the homogeneous one as explicitly described in Ref. [10]), a straightforward extension to the general ϕ^n case was outlined in Ref. [13]. Besides the trivial background term (13) (see Refs. [10] and [43]), one accounts for leading logs through[3]

$$f_\infty(\tau, h) = b^{-d} f_\infty[b^{y_t} (\ln b)^{\hat{y}_t} \tau, b^{y_h} (\ln b)^{\hat{y}_h} h].\tag{34}$$

At the risk of stating the obvious, the same formula was recently claimed in Ref. [46] with the explicit background term but without the backing of theoretical motivation of Luijten and Blöte's RG calculations.[43]

Thermodynamic quantities are derivable as before but with logarithmic corrections:

$$m_\infty(\tau, h) = b^{-d+y_h}(\ln b)^{\hat{y}_h} m_\infty[b^{y_t}(\ln b)^{\hat{y}_t}\tau, b^{y_h}(\ln b)^{\hat{y}_h}h], \tag{35}$$

$$e_\infty(\tau, h) = b^{-d+y_t}(\ln b)^{\hat{y}_t} e_\infty[b^{y_t}(\ln b)^{\hat{y}_t}\tau, b^{y_h}(\ln b)^{\hat{y}_h}h], \tag{36}$$

$$\chi_\infty(\tau, h) = b^{-d+2y_h}(\ln b)^{2\hat{y}_h} \chi_\infty[b^{y_t}(\ln b)^{\hat{y}_t}\tau, b^{y_h}h(\ln b)^{\hat{y}_h}], \tag{37}$$

$$c_\infty(\tau, h) = b^{-d+2y_t}(\ln b)^{2\hat{y}_t} c_\infty[b^{y_t}(\ln b)^{\hat{y}_t}\tau, b^{y_h}h(\ln b)^{\hat{y}_h}]. \tag{38}$$

Setting $b = |h|^{-1/y_h}(\ln|h|)^{-\hat{y}_h/y_h}$ or $b = |\tau|^{-1/y_t}(\ln|\tau|)^{-\hat{y}_t/y_t}$, appropriately, delivers

$$m_\infty(\tau, 0) \sim |\tau|^\beta |\ln|\tau||^{\hat{\beta}} \quad \text{for } T < T_c, \quad \hat{\beta} = \beta\hat{y}_t + \hat{y}_h \tag{39}$$

$$m_\infty(0, h) \sim |h|^{\frac{1}{\delta}}|\ln|h||^{\hat{\delta}}, \quad \hat{\delta} = d\frac{\hat{y}_h}{y_h} \tag{40}$$

$$\chi_\infty(\tau, 0) \sim |\tau|^{-\gamma}|\ln|\tau||^{\hat{\gamma}}, \quad \hat{\gamma} = -\gamma\hat{y}_t + 2\hat{y}_h \tag{41}$$

$$c_\infty(\tau, 0) \sim |\tau|^{-\alpha}|\ln|\tau||^{\hat{\alpha}}, \quad \hat{\alpha} = d\frac{\hat{y}_t}{y_t}. \tag{42}$$

Note that all hatted exponents are by default defined with a positive sign, even when the leading singularity is diverging.

A counterpart formula for the correlation length was not considered in Ref. [42]. In anticipation of superlinearity, we express it as

$$\xi_\infty(\tau, h) = b(\ln b)^{\hat{\digamma}}\xi_\infty[b^{y_t}(\ln b)^{\hat{y}_t}\tau, b^{y_h}(\ln b)^{\hat{y}_h}h]. \tag{43}$$

Here we have used the notation $\hat{\digamma}$ in anticipation of what is to come (as stated, until Ref. [8], we had used the more prosaic symbol $\hat{q}$). We thus have now

$$\xi_\infty(\tau, 0) \sim |\tau|^{-\nu}|\ln|\tau||^{\hat{\nu}}, \quad \hat{\nu} = -\frac{\hat{y}_t}{y_t} + \hat{\digamma}. \tag{44}$$

Eliminating the scaling fields $\hat{y}_t$ and $\hat{y}_h$ in favour of the measurable critical exponents,

$$\hat{y}_t = \frac{\hat{\beta}}{\beta} - \frac{\delta\hat{\delta}}{\beta(1+\delta)} = \frac{2\hat{\beta} - \hat{\gamma}}{2\beta + \gamma} \tag{45}$$

and

$$\hat{y}_h = \frac{\delta\hat{\delta}}{1+\delta} = \frac{\gamma\hat{\beta} + \beta\hat{\gamma}}{2\beta + \gamma}, \tag{46}$$

delivers the scaling relations for logarithmic corrections

$$\hat{\alpha} = 2\hat{\beta} - \hat{\gamma}, \tag{47}$$

$$(\delta - 1)\hat{\beta} = \delta\hat{\delta} - \hat{\gamma}, \tag{48}$$

$$\hat{\nu} = \hat{\digamma} - \frac{\nu(2\hat{\beta} - \hat{\gamma})}{2 - \alpha}. \tag{49}$$

Eqs.(47) and (48) are logarithmic counterparts for the scaling relations (28) and (30), respectively, and inserting them in Eq.(49) delivers Eq.(7). These scaling relations for logarithmic corrections were developed and verified in Refs. [4,5] through very different means (not reliant on phenomenologically modified Widom scaling).

In Ref. [4], one of us stated "relations (47) and (48) but not (49) can be derived starting with a suitably modified phenomenological Widom ansatz." That ansatz was Eq.(34), later put to writing for the 4D Ising model by Aktekin in Ref. [42]. What was missing then was the extension of the "suitably modified phenomenological Widom ansatz" to the correlation sector, namely Eq.(43) and the corresponding homogeneity assumption for the correlation function which reads as[32]

$$g_\infty(\mathbf{x}, \tau, h) = b^{-2x_\phi}(\ln b)^{\hat{\eta}} g_\infty[b^{-1}\mathbf{x}, b^{y_t}(\ln b)^{\hat{y}_t}\tau, b^{y_h}(\ln b)^{\hat{y}_h} h]. \tag{50}$$

That extension, together with Eq.(43), had no basis then because of the evidence that was lacking. Evidence comes from two sources. The first basis is empirical; as documented in Volume III of this series,[3] the three formulae (47), (48) and (49) hold for different universality classes, including the Ising model, $O(N)$ ϕ^4 models, long-range Ising models, m-component spin glasses, percolation, the Yang-Lee edge and lattice animals, all at their respective upper critical dimensions. In each of these cases if $\hat{\digamma}$ were set to zero the scaling relation would fail.

The second basis for Eq.(43) was reported on in Volume IV, some of which we next elaborate upon.

Let us mention here that in Ref. [3], one of the scaling laws among hatted exponents, $\hat{\eta} = \hat{\gamma} - \hat{\nu}(2 - \eta)$, was not correct in the case $\hat{\digamma} \neq 0$. Indeed, we reported in Ref. [32] the correct form,

$$\hat{\eta} = \hat{\gamma} - \hat{\nu}(2 - \eta) + \gamma\frac{\hat{\digamma}}{\nu}, \tag{51}$$

satisfying the condition $g_L(\tau, h, |\mathbf{r}| \to \infty) \to m_L^2(\tau, h)$ which was erroneously relaxed in Ref. [3].

2.3. *The enigmatic nature of $\hat{\digamma}$ and of $\digamma$*

To enable a physical interpretation of $\hat{\digamma}$, we escalate Eqs.(34), (43) and (50) to allow for finite size:

$$f_L(\tau, h, L^{-1}) = b^{-d} f_L[b^{y_t}(\ln b)^{\hat{y}_t}\tau, b^{y_h}(\ln b)^{\hat{y}_h}h, bL^{-1}], \tag{52}$$

$$\xi_L(\tau, h, L^{-1}) = b(\ln b)^{\hat{\digamma}}\xi_L[b^{y_t}(\ln b)^{\hat{y}_t}\tau, b^{y_h}(\ln b)^{\hat{y}_h}h, bL^{-1}], \tag{53}$$

$$g_L(\mathbf{x}, \tau, h, L^{-1}) = b^{-2x_\phi}(\ln b)^{\hat{\eta}}g_L[b^{-1}\mathbf{x}, b^{y_t}(\ln b)^{\hat{y}_t}\tau, b^{y_h}(\ln b)^{\hat{y}_h}h, bL^{-1}]. \tag{54}$$

As described in Ref. [10], this is possible in RG terms because of its local nature — the renormalization constants that apply in the infinite-volume theory, apply for finite volume too.[10] Setting $\tau = h = 0$ in the first two arguments of ξ, and $b = L$ in the third argument (to sit at or in the vicinity of the pseudocritical point $t = 0$), we obtain Eq.(5) as the logarithmic counterpart of Eq.(2). It also enables the QFSS form (11) by expressing the free energy and its derivatives as a function of ξ_∞ and ξ_L instead of τ and L.

Despite the previous widespread dogma that the correlation length cannot exceed the length, this logarithmic superlinearity was explicitly derived by Brézin for the large N limit of the N-vector model (spherical model) for periodic boundary conditions (PBCs). Brézin expected that the picture "is qualitatively unchanged for finite N".[7] In particular, the value $\hat{\digamma} = 1/4$ was explicitly laid out there. This was implicitly verified for PBCs in the four-dimensional Ising model Ref. [9] and explicitly (also for PBCs) in Ref. [47].

The value $\hat{\digamma} = 1/4$ extends to all values of N in the $O(N)$ theory. The N-dependency of partition function zeros and thermodynamic functions were also studied in Ref. [13]. There it was found that leading logarithmic corrections to the finite-size dependency are independent of N in the odd sector — *i.e.*, for Lee-Yang zeros, magnetic susceptibility and related functions. In contrast, Fisher zeros, specific heat and other quantities associated with the even sector were found to be N-dependent. It is now well established on model-specific theoretic and scaling grounds, as well as from numerical simulations that the correlation length of a confined system can exceed its physical length.

And herein lies the enigmatic nature of $\hat{\digamma}$. The three scaling relations (47), (48) and (49) are valid for infinite size but one of them is sourced in finite-size concepts. If hyperscaling (1) is a scaling relation between critical exponents α and ν, its logarithmic counterpart (7) is too and $\hat{\digamma}$

is a critical exponent, just as $\hat{\alpha}$ and $\hat{\nu}$ are. On the other hand $\hat{\varphi}$ is a manifestation of finite-size; it characterises the correlation length through Eq.(53) and does not appear in infinite volume systems where only the exponent ν describes the correlation length in Eq.(25). In that case, $\hat{\varphi}$ is a pseudocritical exponent and Eq.(7) is a finite-size relation. But what then is the status of Eq.(1) if it is to be replaced by Eq.(4)?

In anticipation of a RG foundation, we next relax Eq.(15) to encompass superlinearity as we have done for Eq.(43)

$$\xi_\infty(\tau, h) = b^{\varphi} \xi_\infty[b^{y_t}\tau, b^{y_h} h]. \tag{55}$$

With $\varphi = d/d_{\mathrm{uc}}$ as in Eq.(3) this delivers the modified hyperscaling form (4). Eq.(3) itself was also derived by Brézin for PBCs in Ref. [7] and was numerically verified by Jeff Jones and Peter (A.P.) Young in Ref. [47]. There it was stated that "for free boundary conditions, it seems obvious that even for $d > 4$ the behaviour of the system will be affected when ξ_L becomes of order L, rather than only change when ξ_L becomes of order the much larger length $[L^{5/4}]$".[47] The expectation in Ref. [47] was that standard FSS should apply for free boundary conditions (FBCs) and that the ratio ξ_∞/ξ_L "may also enter" but presumably for "corrections to the scaling terms which involve ξ_∞/L". Ref. [47] ends with the statement "Since FSS for models with FBCs in $d > 4$ is poorly understood, it would be interesting to investigate such models in some detail."

Thus we arrive at a rather puzzling situation. On the one hand, if the new hyperscaling from (4) and its logarithmic counterpart (7) are to stand as scaling relations, φ and $\hat{\varphi}$ would have to be considered as scaling exponents. But since their physical interpretations are finite-size, they are pseudocritical (like λ and ρ to be discussed below). Moreover, since the critical exponents listed by Fisher in Ref. [2] are associated with singularities, they are not burdened by limitations of boundary conditions. So the question arises to what extent φ and $\hat{\varphi}$ are universal — independent of boundaries. This is the main topic of this chapter.

2.4. *Summary*

At this point, we have introduced a very general scaling picture based on a simple assumption of homogeneity in the free energy, correlation function and correlation length. This delivers the standard scaling relations which have been in place since the 1960's and are known to work for $d \leq d_{\mathrm{uc}}$. We then dropped homogeneity and introduced logarithmic corrections to this picture with hatted exponents for each thermodynamic function. In anticipation of the next section, this included an exponent $\hat{\varphi}$ which has no leading-scaling counterpart below the critical dimension. To give physical meaning to $\hat{\varphi}$ we promoted the logarithmically-corrected form to finite-size systems. Thus captured the QFSS form (11) for the critical dimension but not above it. Moreover, at the critical dimension itself, this superlinear scaling behaviour has mathematical and numerical backup, at least for the Ising model and for PBCs, and by extension for the $O(N)$ model, at least in the magnetic (odd) sector. At this point, then, the situation for FBCs is "poorly understood" and puzzling. Universality suggests $\hat{\varphi}$ should manifest there but conservative thinking (that the correlation length cannot exceed the length) suggests it cannot.

We have not, so far, paid attention to the symmetry group of a given underlying Hamiltonian or the structure of the Hamiltonian itself. We have only considered measurable parameters τ or t and h as driving the phase transition. Next, we explore these concepts further as we allow them to take complex values! To do this, we briefly revisit the origins of scaling relations in its broadest context as reported in Volume III.[3] Our purpose here is not to revisit old ground covered in Volume III but to provide context for higher dimensions. For the question arises what is the high dimensional counterpart of $\hat{\varphi}$ and Eq.(6).[e] This will lead us to the new hyperscaling relation Eq.(4). Structuring the RG to achieve this relation opens a whole new area in high dimensional that has been hidden since Fisher first introduced dangerous irrelevant variables (DIVs).

3. The fundamental theory of phase transition as a framework

In addition to thermodynamic and correlation functions, advocates of the partition-function-zero approach have an extra tool at their disposal for the understanding of critical phenomena. One of these was Fa Yueh (Fred) Wu who described them as delivering a "fundamental theory of phase transitions".[49] The zeros approach itself was invented in the 1950's (well before the discovery of the renormalization group) in two papers by Tsung-Dao

---

[e]We mention here an early audacious hypothesis which has not been published ten years ago but was recently proposed to a special issue of the journal Condensed Matter Physics in memory of Ralph.[48]

(T.D.) Lee and Chen Ning Yang.[50,51f] As stated by Wu, "these two papers have profoundly influenced modern-day statistical mechanics."

Partition function zeros are not simple derivatives of the free energy and are usually considered on a model-by-model basis. In terms of the free energy, the partition function itself is given by

$$Z_L(\tau, h) = \exp\left[-\frac{1}{k_B T} F_L(\tau, h)\right].$$ (56)

This is a real function which cannot vanish for finite $F_L(\tau, h)$. When the free energy becomes singular in the infinite-volume limit, however, it can. Yang and Lee's insight was to relax usual (physical) restriction that the external field h be a real parameter. As a complex variable it "opened a new window"[49] by enabling the vanishing of the partition function at values of field now called Lee-Yang zeros. The celebrated Lee-Yang circle theorem, which holds for models with certain symmetries such as the Ising model, states that these zeros occur when h is purely imaginary [on a unit circle in the fugacity $\exp(\beta h)$ plane]. Symmetry demands this is the case whether the system is infinite in size or finite. While interesting, the Lee-Yang theorem is not required for the considerations that follow in this chapter. Still, whatever its shape, the line of Lee-Yang zeros is known as the singular line. In the symmetric phase, for temperatures above the critical one, this line stays away from the real magnetic-field h axis and terminates at what is known as the Yang-Lee edge (this name, too, was proposed by Michael Fisher[52]). As the (real) temperature drops to the critical value, the Yang-Lee edge approaches the real h axis at its critical value $h = 0$. At the critical temperature itself (in infinite volume) the line of zeros pinches the real axis at this point (from the upper and lower halves of the complex h plane) and analytic continuation from $\mathrm{Re}(h) > 0$ to $\mathrm{Re}(h) < 0$ is forbidden. Hence a first-order phase transition is precipitated at $h = 0$. We denote the Yang-Lee edge (in infinite volume) by $r_{\mathrm{YL}}(t)$. Its approach to the real axis is characterised by a power law with the possibility of logarithmic corrections:[9]

$$r_{\mathrm{YL}}(\tau) \sim |\tau|^{\Delta}|\ln|\tau||^{\hat{\Delta}},$$ (57)

where Δ is the gap exponent.[g] We may identify this, and its logarithmic counterpart, from the form of the free energy in Eq.(34) which, on setting

---

[f]A few years later, in 1957, this dynamic pair were awarded the Nobel Prize "for their penetrating investigation of the so-called parity laws which has led to important discoveries regarding the elementary particles."

[g]As stated earlier, Δ represents the "gap" between critical exponents labelling sequential magnetic moments. That it also describes the Yang-Lee "gap" is a fortuitous etymological co-occurrence.

the first argument to a constant is a function of $h|\tau|^{y_h/y_t}|\ln|\tau||^{y_h \hat{y}_t/y_t - \hat{y}_h}$.
Since the partition function holds this functional dependency, its zeros occur
at given values of h, including at $h = r_{\mathrm{YL}}(\tau)$, with

$$\Delta = \frac{y_h}{y_t} = \beta + \gamma = \beta\delta = \frac{\gamma\delta}{\delta - 1} \tag{58}$$

and

$$\hat{\Delta} = \Delta\hat{y}_t - \hat{y}_h = \hat{\beta} - \hat{\gamma}. \tag{59}$$

One may similarly investigate zeros in the complex temperature plane.
These were introduced in 1964 by Michael Fisher in a series of lecture
notes[53] and are called Fisher zeros. While Lee-Yang zeros are those of
the grand canonical partition function, the Fisher zeros are of the canonical partition function only. As for Lee-Yang zeros, these remain away from
the real (temperature) axis if the external magnetic field is non-critical, *i.e.*,
non-zero. As h approaches zero, the zeros approach the real axis and pinch-
ing occurs. However, Fisher zeros are rarely considered in non-vanishing
h. Like Lee-Yang zeros in the complex field plane, Fisher zeros sometimes
lie on curves in the complex temperature plane. However, their distribution is more elaborate than a simple line and may even disperse across
2-dimensional areas (this mostly occurs in circumstances of anisotropy).
As for Lee-Yang zeros, we do not require the precise shape of Fisher zeros
for most of the considerations that follow in this chapter (it was required
for Volume III which concerned logarithmic corrections at low dimensions
as well).

3.1. *Lee-Yang zeros*

Be it for Lee-Yang or Fisher zeros, for finite-size systems the singular line
lacks continuity of its infinite-volume counterpart. Instead one has a discrete bead of zeros. This is clear by writing the partition function as a sum
over all possible configurations of the microscopic degrees of freedom. We
can gather these configurations into a histogram of frequencies of even and
odd macrostates — energy and induced magnetisation, respectively. For
the Ising model, with spins s_i at the sites i of a regular lattice comprising L^d nodes and dL^d links, for example, these are $E = -J\sum_{(i,j)} s_i s_j$ and
$M = \sum_i s_i$, respectively. We set the strength of interaction J to one so that
E can take integer values ranging from $-dL^d$ and dL^d and $M = \sum_i s_i$ takes
integer values from $-L^d$ to L^d. The grand canonical partition function in

the presence of an external magnetic field H is then

$$Z_L(\beta, h) = \frac{1}{\mathcal{N}} \sum_{\{s_i\}} e^{-\beta(-E-HM)} = \frac{1}{\mathcal{N}} \sum_{\{s_i\}} e^{(\beta E + hM)} \ , \tag{60}$$

where $\beta = 1/k_B T$ is the inverse of the Boltzmann constant times the temperature and $h = \beta H$ is the reduced external magnetic field. The factor $\mathcal{N}$ is introduced for normalization purposes and serves no active role in what follows. The partition function may be expressed as

$$Z_L(\beta, h) = \sum_{M=-L^d}^{L^d} \sum_{E=-dL^d}^{dL^d} \rho_L(E, M) e^{\beta E + hM} \ , \tag{61}$$

where the spectral density $\rho_L(E, M)$ denotes the relative weight of configurations having given values of E and M. $Z_L(\beta, h)$ can be written as a polynomial in the fugacity z defined by

$$z = e^{-2h} \ , \tag{62}$$

as

$$Z_L(\beta, h) = z^{-\frac{1}{2} L^d} \sum_{n=0}^{L^d} \rho_{L,\beta}(n) z^n \ , \tag{63}$$

in which

$$\rho_{L,\beta}(n) = \sum_{E=-dL^d}^{dL^d} \rho_L(E, L^d - 2n) e^{\beta E} \tag{64}$$

is an integrated density. That the partition function (63) and the corresponding free energy (as given by its logarithm) are completely analytic for finite L, establishes that no phase transition can occur in a finite-size system. However, as L is allowed to go to infinity phase transitions manifest themselves as points of non–analyticity. For a given value of L and β, Z_L has strictly complex roots z_j. Again, this is consistent with the fact that the free energy is analytic at any real value of the magnetic field H for a finite system; the partition function has no zeros at real H. Complex zeros do, however, exist and we can write the partition in factorized form:

$$Z_L(\beta, h) = A_L(\beta, h) \prod_{j=1}^{L^d} (z - z_j), \tag{65}$$

with $A_L(\beta, h)$ a smooth non-vanishing function. The Lee-Yang theorem states that the zeros lie on the unit circle in the plane of complex variable z, which means they lie on the imaginary h axis.[50,51] As stated, this theorem is not required in what follows.

From conformal invariance, we can just as well operate in the complex h plane and we write the jth zero there as $h_j(\tau)$ where it is to be understood that the actual value is L-dependent. One may then write the finite-size free energy as a sum over Lee-Yang zeros,

$$f_L(\tau,h) = A'_L(\tau,h) L^{-d} \sum_j \ln(h - h_j(\tau)), \tag{66}$$

where $A_L(\tau,h)$ is a normalizing factor which plays no active role in what follows. (Even in the thermodynamic limit, it contributes only to the regular part of the free energy and its derivatives and not to critical behaviour.) We therefore drop it from our considerations. Differentiating twice with respect to field h then gives

$$\chi_L(\tau,h) = -L^{-d} \sum_j \frac{1}{[h - h_j(\tau)]^2} \tag{67}$$

(having dropped a regular additive term). For a finite system, the susceptibility would manifest a singularity if the magnetic field h coincided with a complex Lee-Yang zero $h_j(\tau)$. However, for purely classical systems, complex h is not physical. Only the thermodynamic limit enables the zero impact onto the real axis to precipitate a real phase transition. In this sense complex zeros may be considered as "proto-critical" points[52] — they have the potential to become critical points.

If we assume the critical behaviour is dominated by the first few zeros, or even just the first zero, one arrives at

$$\chi_L(0,0) \sim L^{-d} h_1(0)^{-2}. \tag{68}$$

At this point, we deploy the QFSS form (11) to determine each side of the equation. This amounts to the substitution

$$|\tau| \to L^{\frac{-ϙ}{\nu}} (\ln L)^{-\frac{\hat{ϙ}-\hat{\nu}}{\nu}} \tag{69}$$

in the scaling expressions (41) for the susceptibility and (57) for the Yang-Lee edge to find

$$\chi_L \sim L^{\frac{ϙ\gamma}{\nu}} (\ln L)^{\hat{\gamma}+\gamma\frac{\hat{ϙ}-\hat{\nu}}{\nu}} \tag{70}$$

and

$$h_1 \sim L^{-\frac{ϙ\Delta}{\nu}} (\ln L)^{\hat{\Delta}-\Delta\frac{\hat{ϙ}-\hat{\nu}}{\nu}}. \tag{71}$$

Matching the leading scaling behaviours of these two expressions in Eq.(68) delivers

$$\gamma = 2\Delta - \nu d_{\mathrm{uc}}, \tag{72}$$

which, since hyperscaling in the form (1) always valid for $d = d_{\mathrm{uc}}$, gives $\Delta = \beta + \gamma$ which we already had in (58). This validates our assumption that the critical behaviour of susceptibility is dominated by the scaling of the first zero. Note also that ϙ, or dimensionality d, has dropped out of this equation, so it is valid in all dimensions. Note also that without the ϙ exponent (or if ϙ $= 1$), Eq.(72) would read $\gamma = 2\Delta - \nu d$. This is a standard form for hyperscaling and it fails for $d > d_{\mathrm{uc}}$.

Applying the same considerations to logarithmic corrections gives

$$2\hat{\Delta} = -\hat{\gamma} + (2\Delta - \gamma)\frac{\hat{ϙ} - \hat{\nu}}{\nu}. \tag{73}$$

With Eq.(72), this gives

$$2\hat{\Delta} = d_{\mathrm{uc}}(\hat{ϙ} - \hat{\nu}) - \hat{\gamma}. \tag{74}$$

Note that $\hat{ϙ}$ has not dropped out of this equation. Adhering to the dogma that the correlation length cannot exceed the length of the finite-size system is equivalent to setting $\hat{ϙ} = 0$. If we follow this route, we find the logarithmic correction counterpart of the gap exponents: $2\hat{\Delta} = -(d_{\mathrm{uc}}\hat{\nu} + \hat{\gamma})$. This equation, however, is incorrect.

In the late 1980s and early 1990s, before CERN confirmed the existence of the Higgs boson,[h] the question of triviality of the ϕ^4 model was very much in vogue and logarithmic corrections played a crucial role.[i] Martin Lüscher and Peter Weisz were amongst those to the fore using the RG on this and,

---

[h]Ten years ago (4 July 2012), the ATLAS and CMS collaborations at the Large Hadron Collider (LHC) announced the detection of the Higgs boson, as predicted by the Standard Model of particle physics, and one year later François Englert and Peter Higgs won the Nobel Prize, having predicted the existence of the particle with Robert Brout, decades earlier.

[i]Triviality does not mean that field theories are useless for descriptions of elementary particles and their interactions; they may not be free at all if a finite (albeit large) ultraviolet cutoff is introduced. The lattice achieves precisely this and if it lies well beyond what is experimentally accessible the theory is perfectly valid for as an accurate and mathematically well-defined model of elementary particle interactions, albeit at sufficient energy. Other theories such as pure QED and the standard SU(2) Higgs model are also expected to be trivial.[54] Nonetheless, triviality represents a defect in the theory as the cutoff has to be put in from the outside and cannot be fundamental. The Fields medal was awarded to Hugo Duminil-Copin in 2022 for proving mean-field critical behavior of the 4D Ising model and triviality of the 4D Euclidean scalar quantum field theory.

in a series of papers, addressed scaling laws and triviality bounds in lattice ϕ^4 theory.[54–56] Their review drew on earlier perturbation-theory results which were "perhaps not as well-known to the lattice gauge community as they deserve to be".[44,57] From these results, it was known that $\hat{\nu} = 1/6$ and $\hat{\gamma} = 1/3$ so $\hat{\digamma} = 0$ would force $\hat{\Delta} = -1/2$ and standard FSS would force $h_1 \sim L^{-\frac{\hat{\Delta}}{\nu}}(\ln L)^{\hat{\Delta}+\Delta\frac{\hat{\digamma}}{\nu}}$. For the Ising universality class this would give $h_1 \sim L^{-3}(\ln L)^0$. The Q-alternative that $\hat{\digamma} = 1/4$ leads to $\hat{\Delta} = 0$ so that $h_1 \sim L^{-3}(\ln L)^{-1/4}$. A negative exponent for the log term, compatible with this latter prediction, was confirmed in Ref. [10] using simulations of the 4D Ising model with $L = 8$ to 24 using the Swendsen–Wang cluster algorithm.

Thus we conclude that scaling relations for logarithmic corrections from a modified Widom hypothesis adhering to the non-superlinear dogma are not correct. Instead, superlinearity as identified forty years ago by Brézin is supported by numerics. Of course Brézin's calculations and the numerics contained in Ref. [10] are for the Ising model in four dimensions with PBCs only. We will broaden these to far more general circumstances in what follows.

3.2. *Fisher zeros*

A similar approach was used in Ref. [9]. Written in terms of the Fisher zeros, the canonical partition function is

$$Z_L(\beta) = \sum_{M=-dL^d}^{dL^d} \rho_L(E)e^{\beta E}, \tag{75}$$

where the spectral density $\rho_L(E)$ counts configurations with given values of E only. The partition function is a polynomial in $y = \exp{(-\beta)}$ and the counterpart of Eq.(65) is

$$Z_L(\beta) = A_L(\beta) \prod_{j=1}^{dL^d}(y - y_j). \tag{76}$$

The free energy may then be written as a sum over Fisher zeros so that

$$f_L(\tau) = A'_L(\tau)L^{-d} \sum_j \ln(\tau - \tau_j), \tag{77}$$

having again used conformal invariance to switch to the complex τ plane where the zeros are labeled τ_j and implicitly depend on L. Differentiating appropriately with respect to temperature, one may write the specific heat in terms of Fisher zeros as

$$c_L(t) = -L^{-d}\sum_j \frac{1}{(\tau-\tau_j)^2}. \tag{78}$$

Again setting $\tau = 0$ and taking the leading scaling to come from the first zero, one arrives at

$$c_L(0) \sim L^{-d}\tau_1^{-2}, \tag{79}$$

where τ_1 is a measure of the difference between the first zero and the critical point.

QFSS in the form (69) applied to c gives

$$c_L(0) \sim L^{\frac{ϙ\alpha}{\nu}}(\ln L)^{\hat\alpha+\frac{\alpha(\hat{ϙ}-\hat\nu)}{\nu}}. \tag{80}$$

To determine QFSS of the Fisher zeros we again appeal to the scaling ratio $x = \xi_L(0)/\xi_\infty(\tau)$ in Eq.(11). Expressing the partition function in terms of x, one finds $Z(\tau) = 0$ when $\tau = \tau_1$ with

$$\tau_1 \sim L^{-\frac{ϙ}{\nu}}(\ln L)^{\frac{\hat\nu-\hat{ϙ}}{\nu}}. \tag{81}$$

Combining Eqs.(79) and (81) through Eq.(79), dimensionality d again drops out of the leading scaling relations to recover $\nu d_{\mathrm{uc}} = 2-\alpha$, and one recovers the scaling relation (7): $\hat\alpha = d(\hat{ϙ} - \hat\nu)$.[j]

Again, ignoring $ϙ$ would give the standard hyperscaling form $\nu d = 2-\alpha$, which fails above the upper critical dimension.

Ignoring $\hat{ϙ}$ or setting $\hat{ϙ} = 0$ in Eq.(81), would deliver $L^2\tau_1(L) \sim (\ln L)^{1/3}$ for Eq.(79) when mean-field values of α and ν are inserted. Incorporating $\hat{ϙ} = 1/4$, on the other hand, delivers $L^2\tau_1(L) \sim (\ln L)^{-1/6}$. Indeed a negative logarithmic exponent compatible with this was observed for the 4D Ising model in Ref. [9]. This establishes a positive (non-zero) value of $\hat{ϙ}$

The inverse Eq.(79) above was used in Ref. [9] to express the lowest-lying Fisher zeros in terms of specific heat:

$$\tau_1 \sim \frac{1}{\sqrt{L^d c_v}}. \tag{82}$$

There, this was used as an alternative approach to finding the FSS of Fisher zeros — an approach not reliant on the QFSS form (11). With $\alpha = 0$, and the logarithmic exponent $\hat\alpha = 1/3$ having again been provided by

---

[j]This is not the full story, however, and a subtlety arises in circumstances where the leading exponent α vanishes and the Fisher zeros impact at an angle of $\pi/4$. This happens in the 2D Ising model, for example. We refer the reader to Ref. [5] for details.

Lüscher, Weisz and others,[44,54–56] this also predicts that the first zero scales as $\tau_1 \sim L^{-2}(\ln L)^{-1/6}$, as verified.[3]

A similar but more sophisticated approach was recently used by Aydin Deger and Christian Flindt.[58–61] They used fluctuations of the total energy and the magnetisation to extract partition function zeros in systems with surprisingly small lattices (see also Ref. [62] where this becomes a strategy to avoid resort to large-scale simulations which cost a lot in terms of carbon footprint). The extra degree of sophistication provided by Deger and Flindt was to combine cumulants of different orders (not just the second-order cumulants used above). Moreover, with their approach, "critical exponents can be determined even if the system is away from the phase transition, for example at a high temperature." Finite-size scaling in high dimensions is addressed in Ref. [60]. The methods introduced by Deger and Flindt are powerful ways to extract partition function zeros from finite-size systems solely using fluctuations of thermodynamic observables and do not require prior knowledge of the partition function itself. Thus their approach opens new ways to access zeros numerically and in experiments.

Indeed, partition function zeros have found relevance for experiments only recently. Long thought to be mathematical constructs without physical realization, in 2015 manifestations of imaginary fields were observed in magnetic resonance experiments performed on the spins of a molecule in Ref. [63]. The experiments followed a theoretical proposal made by Bo-Bo Wei and Ren-Bao Liu a few years earlier.[64] They had considered an Ising spin bath and found an equivalence between its partition function with complex field and the quantum coherence of a probe spin coupled to the spin bath; "the times at which the quantum coherence reaches zero are equivalent to the complex fields at which the partition function vanishes—that is, the fields that produce the Lee-Yang zeros". This was nicely verified in experiments by Xinhua Peng and her colleagues Ref. [63]. See Ref. [65] for a non-technical summary of these important theoretical and experimental results.[63,64]

3.3. *Pseudocriticality, shifting and rounding*

In the discussion around Eqs.(67) and (68), we purposefully relaxed notation. In particular, we did not pay heed to any difference between the real and imaginary parts of the partition function zeros. If our focus is on Lee-Yang zeros, in situations where the Lee-Yang theorem holds, the zeros necessarily lie on a singular line which is on the imaginary axis. *I.e.*, the real part of the Lee-Yang zeros is zero for any lattice size. In this case, there is no ambiguity and $h_1(\tau)$ refers to the imaginary part of the first Lee-Yang zero.

In the case of Fisher zeros, however, the situation is slightly more complicated. Here τ_1 has both real and imaginary parts which depend upon the system size L. Let us suppose the leading FSS behaviour for the real part is $\mathrm{Re}[\tau_1(L)] \sim |\tau|^{-1/\nu_{\mathrm{real}}}$ and that for the imaginary part is as $\mathrm{Im}[\tau_1(L)] \sim |\tau|^{-1/\nu_{\mathrm{imag}}}$. The partition function zero then scales as the slower of these two so that $\nu = \min(\nu_{\mathrm{real}}, \nu_{\mathrm{imag}})$. The Fisher zeros impact onto the real axis at an angle given by $\tan\phi \approx \mathrm{Im}[\tau_1(L)]/\mathrm{Re}[\tau_1(L)] \sim L^{\nu_{\mathrm{real}}-\nu_{\mathrm{imag}}}$. If $\nu_{\mathrm{real}} < \nu_{\mathrm{imag}}$, the impact angle would be zero in the infinite-volume limit, meaning a spread of singularities instead of a single critical point. If $\nu_{\mathrm{real}} > \nu_{\mathrm{imag}}$, on the other hand, the zeros impact vertically. This means there is a symmetry between the ordered and disordered phases and the specific heat amplitudes on either side of the critical point have to coincide. Indeed, this happens in the 2D Ising model because of its self-dual property. If $\nu_{\mathrm{real}} = \nu_{\mathrm{imag}}$, any angle of impact is possible. For a full discussion of these matters, see Refs. [3,66].

Therefore, in most circumstances the real part of the first Fisher zero scales to the infinite-volume critical point in the same way as the imaginary part approaches zero. The only circumstance where that might not be expected to happen is when the specific-heat amplitude ratio (a universal quantity) is one.

Moreover, Eq.(79) suggests that specific heat peaks when τ_1 is smallest, *i.e.*, when the temperature is close to the real part of the first Fisher zero. This means that the real part of the first zero is a pseudocritical point and this is shifted away from the critical point by an amount proportional to $L^{-\nu_{\mathrm{real}}}$, to leading order. The location of the specific heat peak and the susceptibility peak are other pseudocritical points. To encapsulate the more general nature of pseudocritical points, we follow traditional notation for the so-called shift exponent and denote it by λ (not to be confused with the generic critical exponent λ used in the quotation by Fisher in Section 2).

For a system of linear extent L, then, the pseudocritical point $\tau_L = T_L/T_c - 1$ scales to leading order potentially with logarithms as

$$|\tau|_L \sim L^{\lambda}(\ln L)^{\hat{\lambda}}, \tag{83}$$

where

$$\lambda = \frac{1}{\nu} \tag{84}$$

provided the specific heat amplitude ratio is not one.

As stated, we do not wish to revisit the old ground covered in Volume III and Volume IV of this series. Instead, we simply summarise how these critical exponents are related and refer the reader to Refs. [3,21] for details. Suffice to say that the scaling relation associated with the logarithmic counterpart to the shift exponent λ is

$$\hat{\lambda} = -\hat{y}_t = \frac{\hat{\nu} - \hat{\varrho}}{\nu}. \tag{85}$$

Pseudocriticality in this context can be defined in a number of ways and may even depend on the quantity Q itself. Strictly speaking, we should use a notation such as τ_L^Q to enforce this point. In that case, τ_L^χ would vanish at the value of T where the magnetic susceptibility reaches its peak value. Likewise, τ_L^ξ would vanish at the value of T where the correlation length peaks. The correlation function doesn't have a peak, of course, and one way to identify it may be as the value of T at which it becomes pure power law, possibly with logarithmic corrections. Again, if we are to be strict, we should also introduce different notations for different reduced temperatures for different functions. Thus τ^χ would represent $T/T_L^\chi - 1$ and so on. Thus there is no ambiguity in the definition of pseudocritical points or the reduced temperatures for finite-size systems. We find the notation τ^Q and τ_L^Q with different Q's unwieldy, however, so we use τ and τ_L throughout. The precise definition should be clear from the context. Regardless of the definition of τ and T_L, they always revert to τ and T_c in the thermodynamic limit.

Finally, we also mention the rounding exponent θ and its logarithmic counterpart $\hat{\theta}$. This is associated with the smoothening out of an infinite-volume divergence into a finite-volume peak. One (rather arbitrary) measurement of rounding is the width of a particular curve at half of its maximum height. Thus the rounding associated with the susceptibility, for example, is the width of the susceptibility curve when susceptibility attains half its maximum value for finite size. We write

$$\Delta T \sim L^{-\theta} (\ln L)^{\hat{\theta}} \tag{86}$$

to leading order in logarithms.

3.4. *Summary*

At this stage of our considerations, we still have not paid attention to the details of any given underlying Hamiltonian. We have not even considered symmetries that lead to the Lee-Yang theorem or Fisher circles. We have limited our presentation to circumstances where the partition function zeros fall on a singular line but even that is not a requirement for our considerations. See Ref. [67] for circumstances where the zeros are not restricted to a curve in the complex plane and/or come in degenerate sets.

To take this further, we wish next to get to the heart of the reasons for superlinear correlation length and the origins of the φ and $\hat{\varphi}$ exponents. The issues we wish to address are, then, the correlation sector, hyperscaling, and finite-size scaling (FSS) in very general renormalization-group terms. Our contribution to Volume III sets the framework: if $\hat{\varphi}$ is required for self-consistency of FSS with logarithmic corrections, including at the critical dimension, what is the source and role of its counterpart φ in higher dimensions? This is where RG comes into play. Until the work presented in Volume IV of this book series, dangerous irrelevant variables (DIV) had been considered only to apply to the free energy of a given model and related thermodynamic functions, which otherwise might fail to match the valid predictions of mean-field theory. DIVs were believed not to manifest in the correlation sector which appeared not to fail. The other two mechanisms, hyperscaling and FSS, were considered to fail in high dimensions. Each of these three beliefs — the non-failure of the correlation sector and the failure of both FSS and hyperscaling — were overturned in Volume IV, in a move that endowed the renormalization group with a more complete degree of encompassment.[21] Here we extend those considerations to ϕ^n theories and update them with the latest relevant literature.

Our purpose in this section is not to re-derive the full details of scaling relations for log corrections. We refer to Refs. [3,32] for a more comprehensive treatment. Instead, our intent here is to give a pedagogic and self-contained account of superlinearity and the validity of QFSS — a framework around which a self-consistent theory must be built. In the above considerations we have seen how critical behaviour is driven by the first few Lee-Yang or Fisher zero. The gross approximations of dropping the summations in Eqs.(67) and (79) are justified posteriori — it works. For a more extensive exposition, higher-order zeros and the density of zeros along the singular line should be accounted for. We refer the reader to Refs. [4,5] for such an account.

4. Ginzburg-Landau theory (mean-field)

In Ginzburg-Landau theory, we consider physical systems described in thermal equilibrium by the partition function

$$Z = \int D\phi\, e^{-F[\phi]}, \tag{87}$$

where the functional

$$F[\phi] = \int d^d x\, f(\phi, \boldsymbol{\nabla}\phi) \tag{88}$$

is a free energy (or action) integrated over all space and

$$f(\phi, \boldsymbol{\nabla}\phi) = \tfrac{1}{2}r\phi^2(\mathbf{x}) + \tfrac{1}{3}u_3\phi^3(\mathbf{x}) + \tfrac{1}{4}u_4\phi^4(\mathbf{x}) + \tfrac{1}{6}u_6\phi^6(\mathbf{x}) - h\phi(\mathbf{x}) + \tfrac{1}{2}|\boldsymbol{\nabla}\phi|^2 \tag{89}$$

is a free energy density. In presenting the free energy density as a power expansion of the order parameter and its derivatives, this expression captures the essence of a multitude of models in statistical physics and, as such, can be applied to many different systems. In reference to any specific system, such as the Ising model itself, the Ginzburg-Landau-Wilson theory may be considered phenomenological. According to the Oxford Learners' Dictionary, the word "phenomenological" pertains to "the branch of philosophy that deals with what you see, hear, feel, *etc.* in contrast to what may actually be real or true about the world." In this sense, it is not fundamental. However, describing ϕ^4 theory in this way misses the point. The ϕ^4 theory removes details such as the quantum nature of the spins behind the Ising model or the precise details of interaction strengths. It strips these back to the bare minimum in terms of dimension and symmetry group that is required for a phase transition in the same universality class. In the case of the Ising model, there is even an (almost) exact mathematical transformation (called the Hubbard-Stratonovich transformation, see *e.g.* Ref. [68]) that maps the Ising model onto the corresponding field theory ("almost" since a $\ln\cosh$ term is expanded to quartic order). Thus, in moving to the Ginzburg-Landau-Wilson model we are not losing any fundamental aspect of the theory — it is not a mere simplification of a specific model, it captures the real and true essence of what is dimensionality, symmetry and boundary conditions without worrying about non-universal details.

To avoid unnecessary clutter we present this chapter in terms of a scalar matter field with $O(1)$ symmetry rather than in terms of vector fields. The extension of the considerations presented in this chapter to the general N case is similarly obvious. These $O(N)$ theories involve higher values of N

and we refer the reader to Ref. [13] for a finite-size scaling theory *at* the upper critical dimension.

The coefficient r in Eq.(89) is a reparameterisation of the reduced temperature τ we had earlier. We take it as positive in the disordered phase and negative in the ordered one. This is required by consistency to have a vanishing order-parameter in the disordered phase as we will see below. In the case of percolation, for example, there is no temperature and r is anyway preferred (over τ). We denote by u_n the coefficient of the highest power of ϕ in the above Lagrangian. The case $n = 3$ refers to percolation and the case $n = 4$ with $u_3 = 0$ is the standard Ginzburg-Landau-Wilson model. As stated, it maps (almost exactly) to the Ising model. The case $n = 6$ is of a tricritical point which marks the singular behaviour at the end of a line of first-order phase transitions. An example of this is found in the Blume-Capel model.[62,69–71] Thus, despite trimming it back to bare essentials, the ϕ^n model indeed covers a range of universality classes.

A simplified version called the Ginzburg-Landau theory does not take into account fluctuations of the order parameter close to the critical point and in this sense, it is a mean-field theory where only averages count (an even simpler version in which the order parameter is spatially uniform is called the Landau theory). Nonetheless, the theory is accurate in most of its predictions when the system is connected enough and this is the case for high-dimensional systems. To access it, we identify the field configurations $\phi_0(\mathbf{x})$ which have the highest weight,

$$\left.\frac{\delta F}{\delta \phi}\right|_{\phi_0(\mathbf{x})} = 0, \tag{90}$$

from which

$$\frac{\partial f}{\partial \phi} - \mathbf{\nabla} \cdot \frac{\partial f}{\partial (\mathbf{\nabla}\phi)} = 0, \quad \text{at} \quad \phi = \phi_0(\mathbf{x}). \tag{91}$$

The gradient term in Eq.(89) does not have an associated coefficient because it has been absorbed into the other coefficients to render the term dimensionless once integrated over space. Therefore, the matter-field scaling dimension, x_ϕ, is

$$x_\phi = \frac{d}{2} - 1. \tag{92}$$

Compare this to Eq.(31), which was derived from power counting. We see that the η exponent vanishes for mean-field theory.

We consider the ϕ^n, then, in its generic form:

$$f(\phi, \mathbf{\nabla}\phi) = \tfrac{1}{2}r\phi^2(\mathbf{x}) + \tfrac{1}{n}u_n\phi^n(\mathbf{x}) - h\phi(\mathbf{x}) + \tfrac{1}{2}|\mathbf{\nabla}\phi|^2. \tag{93}$$

Dropping the gradient term for an infinite homogeneous system, Eq.(90) gives

$$\phi_0(r + u_n\phi_0^{n-2}) = h. \tag{94}$$

If $h = 0$ we identify the order parameter m_∞ as ϕ_0 so that

$$m_\infty(\tau) = (-r/u_n)^{\frac{1}{n-2}}, \qquad T < T_c, \tag{95}$$

$$m_\infty(\tau) = 0, \qquad T > T_c. \tag{96}$$

From this we extract the critical exponent

$$\beta_{\mathrm{MFT}} = \frac{1}{n-2}. \tag{97}$$

and magnetisation amplitude in the ordered phase, as used in Eq.(21), as then $B^- = (u_n)^{-1/(n-2)}$. The magnetic field dependency of the order parameter at the critical temperature $r = 0$ also comes from Eq.(94) and is

$$m_\infty(h) = \mathrm{sgn}\,(h)\,(|h|/u_n)^{\frac{1}{n-1}}, \qquad T = T_c, \tag{98}$$

from which, comparing to Eq.(20),

$$\delta_{\mathrm{MFT}} = n - 1 \tag{99}$$

together with the amplitude $D_c = (u_n)^{-1/(n-1)}$.

The second derivative of Eq.(94) wrt h gives the susceptibility

$$\chi_\infty = [r + (n-1)u_n\phi_0^{n-2}]^{-1}. \tag{100}$$

Eqs.(95) and (96) then give

$$\chi_\infty(\tau) = [(n-2)(-r)]^{-1}, \qquad T < T_c, \tag{101}$$

$$\chi_\infty(\tau) = r^{-1}, \qquad T > T_c, \tag{102}$$

for the two phases. Both of these deliver the same critical exponent [see Eq.(22)]

$$\gamma_{\mathrm{MFT}} = 1, \tag{103}$$

and the associated amplitudes are $\Gamma^- = (n-2)^{-1}$ and $\Gamma^+ = 1$.

The free energy is given by inserting the equilibrium order parameter (95) and (96) in the expansion (93), so that

$$f_\infty(\tau) = \left(\frac{1}{n} - \frac{1}{2}\right) u_n^{\frac{2}{2-n}}(-r)^{\frac{n}{n-2}}, \qquad T < T_c, \tag{104}$$

$$f_\infty(\tau) = 0, \qquad T > T_c. \tag{105}$$

The second temperature derivative gives the specific heat exponent from Eq.(23) as

$$c_\infty(\tau) = \frac{1}{2-n} u_n^{\frac{2}{2-n}} (-r)^{\frac{4-n}{n-2}}, \qquad T < T_c, \tag{106}$$

$$c_\infty(\tau) = 0, \qquad T > T_c. \tag{107}$$

Thus a jump appears at the phase transition and the exponent is associated with the low-temperature regime only (for this, the mean-field solution). Notwithstanding this, we write,

$$\alpha_{\mathrm{MFT}} = \frac{n-4}{n-2}, \tag{108}$$

with the amplitude in this regime given by

$$(A^-/\alpha_{\mathrm{MFT}}) = (u_n)^{-2/(n-2)}/(2-n). \tag{109}$$

For the correlations, one has to reinstate the gradient term to Eq.(93). The Euler-Lagrange equation (91) then leads to

$$r\phi(\mathbf{x}) - u_n \phi^{n-1}(\mathbf{x}) - \nabla^2 \phi(\mathbf{x}) = h. \tag{110}$$

The space dependency of the correlation function can be extracted from the order parameter profile when a localized magnetic field $h_0 \delta(\mathbf{x})$ is applied at the origin. At criticality $r = h = 0$, and $\phi(\mathbf{x})$ can be considered small enough to neglect the non-linear term. Outside the origin, this leads to a Laplace equation which happens to be independent of the value of n at this level of approximation. One finds (assuming isotropy)

$$\nabla^2 \phi(\mathbf{x}) = \frac{1}{|\mathbf{x}|^{d-1}} \frac{d}{d|\mathbf{x}|} \left(|\mathbf{x}|^{d-1} \frac{d\phi(\mathbf{x})}{d|\mathbf{x}|} \right) = 0. \tag{111}$$

The solution takes the form

$$g(\mathbf{x}) \sim \frac{1}{|\mathbf{x}|^{d-2}}. \tag{112}$$

This is consistent with the critical exponent

$$\eta_{\mathrm{MFT}} = 0. \tag{113}$$

Again, this does not depend on n. Beyond the critical temperature, [say above T_c, since we are still neglecting the ϕ^{n-1} term in Eq.(110)][k] the equation that has to be solved is

$$r\phi(\mathbf{x}) - \nabla^2 \phi(\mathbf{x}) = h. \tag{114}$$

---

[k]Below the critical temperature we may assume the same τ dependency for the correlation length (an argument that can be made more rigorous).[72]

The only one at our disposal in this, the infinite-volume, thermodynamic, limit is the correlation length. For this reason, we identity the correlation lengths $\xi(\tau, h = 0)$ and $\xi(\tau = 0, h)$

$$\xi(\tau, h = 0) \sim 1/|r|^{1/2}, \qquad \xi(\tau = 0, h) \sim |\phi/h|^{1/2}, \tag{115}$$

for the two different phases. With $\phi \sim h^{1/\delta}$, both exponents of the correlation length follow:

$$\nu_{\mathrm{MFT}} = 1/2, \qquad \nu_{\mathrm{c\,MFT}} = \frac{\delta - 1}{2\delta} = \frac{n - 2}{2(n - 1)}. \tag{116}$$

For convenience, we collect the mean-field exponents for the Ising, percolation and tricriticality universality classes in Table 1.

Table 1. Critical exponents for the Gaussian model and mean-field critical exponents for percolation, Ising magnets (the SAW has the same exponents) and for tricriticality as well as for the generic Landau model. RG eigenvalues are $y_t = 2$, $y_h = d/2 + 1$ and $y_u = n + (2 - n)d/2$. The latter is negative when $d > d_{\mathrm{uc}}$. Despite this irrelevancy of the field u, only γ, ν and η match the Gaussian prediction.

Model	ϕ^n	α	β	δ	ν_c	γ	ν	η	d_{uc}
GFP	ϕ^2	$2 - \frac{d}{2}$	$\frac{d-2}{4}$	$\frac{d+2}{d-2}$	$\frac{2}{d+2}$	1	$\frac{1}{2}$	0	$\frac{2n}{n-2}$
Percolation	ϕ^3	-1	1	2	$\frac{1}{4}$	1	$\frac{1}{2}$	0	6
Magnets, SAW	ϕ^4	0	$\frac{1}{2}$	3	$\frac{1}{3}$	1	$\frac{1}{2}$	0	4
Tricriticality	ϕ^6	$\frac{1}{2}$	$\frac{1}{4}$	5	$\frac{2}{5}$	1	$\frac{1}{2}$	0	3
Landau	ϕ^n	$\frac{n-4}{n-2}$	$\frac{1}{n-2}$	$n - 1$	$\frac{n-2}{2(n-1)}$	1	$\frac{1}{2}$	0	$\frac{2n}{n-2}$

Finally, we reflect on where the above considerations are valid. Fluctuations are measured by the susceptibility, of course, and this is an integral over space of the correlation function:

$$\chi \simeq \int d^d x\, g(\mathbf{x}) \sim |\tau|^{-\gamma_{\mathrm{MFT}}}. \tag{117}$$

The square of the magnetisation inside the correlation volume, on the other hand, is

$$\xi^d m_\infty^2 \sim |\tau|^{-d\nu_{\mathrm{MFT}} + 2\beta_{\mathrm{MFT}}}. \tag{118}$$

These two quantities have the same dimensions and, comparing them, fluctuations are relatively weak if $d \geq d_{\mathrm{uc}}$ where

$$d_{\mathrm{uc}} = \frac{2\beta_{\mathrm{MFT}} + \gamma_{\mathrm{MFT}}}{\nu_{\mathrm{MFT}}}. \tag{119}$$

The inequality $d \geq d_{\mathrm{uc}}$ is the Ginzburg criterion — a neat indicator of where order parameter fluctuations can be neglected and where mean-field theory kicks in. Inserting Eqs.(97), (103) and (116) for the mean-field exponents, we then get

$$d_{\mathrm{uc}} = \frac{2n}{n-2}. \tag{120}$$

The demarcation point d_{uc} is the upper critical dimension and its values for different models are given in Table 1.

5. The Gaussian fixed point: its apparent sufficiency for the correlation sector and its insufficiency for the free-energy sector

The dimensionlessness of the partition function in Eq.(87) dictates the dimensionality of each individual term in the free energy density (89) or (93). From this we obtain the scaling dimension of the matter field as Eq.(92) and

$$y_t + 2x_\phi = d, \; y_t = 2, \tag{121}$$

$$y_h + x_\phi = d, \; y_h = \frac{d}{2} + 1, \tag{122}$$

$$y_u + nx_\phi = d, \; y_u = \frac{d}{2}(2 - n) + n. \tag{123}$$

Here, and henceforth, we use u for u_n and y_u for y_{u_n}). The renormalization flow $\tau' = b^{y_t}\tau$, $h' = b^{y_h}h$, $u' = b^{y_u}u$ is controlled by these RG eigenvalues, and $\tau = h = u = 0$ is identified as the Gaussian Fixed Point (GFP). This is because there the partition function (87) becomes a Gaussian integral,

$$Z = \int D\phi \, e^{-\int d^d x \, \frac{1}{2}|\nabla\phi|^2}. \tag{124}$$

The additional scaling field u takes us beyond the simple homogeneous form (13) to

$$f_\infty^{\mathrm{sing}}(\tau, h, u) = b^{-d} f_\infty^{\mathrm{sing}}(b^{y_t}\tau, b^{y_h}h, b^{y_u}u). \tag{125}$$

Likewise, the correlation function and correlation length take the forms

$$g_\infty^{\mathrm{sing}}(\mathbf{x}, \tau, h, u) = b^{-2x_\phi} g_\infty^{\mathrm{sing}}(b^{-1}\mathbf{x}, b^{y_t}\tau, b^{y_h}h, b^{y_u}u) \tag{126}$$

and

$$\xi_\infty^{\mathrm{sing}}(\tau, h, u) = b\xi_\infty^{\mathrm{sing}}(b^{y_t}\tau, b^{y_h}h, b^{y_u}u), \tag{127}$$

respectively.

The RG eigenvalues y_t and y_h in Eqs.(121) and (121) are positive so their associated scaling fields are relevant in the sense that they drive the system away from the fixed point ($\tau = h = u = 0$) as the RG is applied. *I.e.*, under rescaling by a factor $b > 0$, they grow as $\tau \to \tau'$ and $h \to h'$ with

$$\tau' = b^2 \tau, \quad \text{and} \quad h' = b^{\frac{d}{2}+1} h. \tag{128}$$

While the scaling field u can also have a positive RG eigenvalue in Eq.(123), this is only the case when the Ginzburg criterion (120) fails. However, when the Ginzburg criterion holds, $y_u < 0$ and starting from any non-zero value of u drives the system back to the critical point. For this reason, u is said to be irrelevant above the upper critical dimension — its inclusion does not affect the universality class (critical exponents) of the model. Precisely at d_{uc}, one has the marginal situation where multiplicative logarithmic corrections to scaling arise.

Therefore, naively (or without paying attention to fifty years of literature), one might expect the predictions from RG at the GFP to match those coming from Landau mean-field theory of the previous section and we also list them in Table 1 for ease of comparison. We extract these by inserting the scaling dimension (92) and the RG eigenvalues (121) and (122) into Eqs.(20)–(31).

There are a number of interesting observations to make here. Firstly the exponents γ, ν, η and the critical dimension all match the correct values coming from MFT. Secondly, all of these values are independent of n. This, of course, is because the scaling dimensions (121) and (122) are independent of n and, while the n-dependency lives only in y_u [Eq.(123)], that is irrelevant for the GFP above the upper critical dimension. The critical exponents listed for percolation, the Ising model and tricriticality have numerous verifications in the literature, confirming that the Landau exponents and not the GFP predictions are the correct ones.

All critical exponents listed in the table obey the scaling relations (28)–(30) as well as (32), derived from simple homogeneous assumptions outlined in Sec. 2.1. The hyperscaling relation (27) only holds for the GFP. Although it holds precisely at the upper critical dimension for percolation, Ising magnets, tricriticality and the general Landau scheme, it does not hold for general $d > d_{\mathrm{uc}}$.

The key to unravelling why the GFP fails to match Landau theory above d_{uc}, and why hyperscaling fails in Landau theory there too, is the observation that the wrong exponents α, β and δ all come from derivatives of the free energy, while the correct γ, ν and η are each associated with correlations. This is why we distinguish between the "free energy sector" and "correlation sector" in the introduction to this chapter and the title of this section.

There are two exceptions to the rule that are extremely informative. Firstly, although the susceptibility belongs to the free energy sector, the GFP delivers the correct mean-field value of the critical exponent for it. This is because, besides being a direct derivative of the free energy, the susceptibility can also be extracted by integrating the correlation function. Therefore it also belongs to the correlation sector. This suggests the robustness of the correlation sector as delivering the correct Landau exponents from the GFP. However, the GFP value for $\nu_{\rm c}$ is $\nu_{\rm c\,G} = 2/(d+2)$ from inserting Eq.(122) for y_h into Eq.(26). This does not agree with the mean-field value in Eq.(116) except when $d = d_{\rm uc}$ (where all exponents coincide). This suggests that something is also wrong in the correlation sector! This observation, along with the search for the leading counterpart of $\hat{\varphi}$ guides the way to open new insights into high dimensions that lay hidden since the inception of RG.

6. Dangerous irrelevancy: Rescuing the free-energy sector in infinite volume and attempts at resuscitation in finite volume

At this point, we have seen that, while the self-interacting field u is irrelevant above the upper critical dimension, the GFP is only partially successful in its delivery of critical exponents there. Moreover, that success lies in the correlation sector only. Also, while hyperscaling matches the GFP, it fails for Landau theory when the Ginzburg criterion holds. Next, we follow Fisher and seemingly rescue the situation for the thermodynamic limit before we encounter another obstacle in finite systems.

6.1. *Dangerous irrelevant variables for thermodynamic functions at infinite volume*

Gathering the seemingly innocuous amplitudes of Eqs.(95), (98) and (106), namely

$$B^- = u^{-\frac{1}{n-2}}, \tag{129}$$

$$D_c = u^{-\frac{1}{n-1}}, \tag{130}$$

$$\frac{A^-}{\alpha_{\mathrm{MFT}}} = \frac{1}{2-n} u^{-\frac{2}{n-2}}, \tag{131}$$

we spot a danger: they are each singular when $u \to 0$! Michael Fisher highlighted this in the 1980's[73] although he may have noticed it ten years earlier.[74] Therefore, although "irrelevant" in the sense of RG flow, the field u is *dangerous* for the amplitudes. The amplitudes of the usual quantities in the correlation sector are not u-dependent and do not face the same danger. An exception to this observation is ξ_c, the h−dependent correlation length at T_c. This quantity is seldom discussed, however, so, as many others have done, we leave it aside for the moment. Suffice it to say for now that u is a *dangerous irrelevant variable* (DIV) for the free-energy sector.

If we define the exponents appearing in Eqs.(129)–(131) as

$$\kappa = \frac{1}{n-2}, \quad \lambda = \frac{1}{n-1}, \quad \mu = \frac{2}{n-2}, \tag{132}$$

we may write the free-energy derivatives as

$$m_\infty(\tau < 0, h = 0, u) \sim |\tau|^\beta u^{-\kappa}, \tag{133}$$

$$m_\infty(\tau = 0, h, u) \sim |h|^{1/\delta} u^{-\lambda}, \tag{134}$$

$$c_\infty(\tau, h = 0, u) \sim |\tau|^{-\alpha} u^{-\mu}. \tag{135}$$

Eqs.(16)–(19) have clearly to be modified to take the DIVs into account and we write

$$m_\infty(\tau, 0, u) \overset{u \to 0}{=} b^{-d+y_h-\kappa y_u} u^{-\kappa} \mathcal{M}^-(b^{y_t}\tau, 0), \tag{136}$$

$$m_\infty(0, h, u) \overset{u \to 0}{=} b^{-d+y_h-\lambda y_u} u^{-\lambda} \mathcal{M}_c(0, b^{y_h}h), \tag{137}$$

$$c_\infty(\tau, 0, u) \overset{u \to 0}{=} b^{-d+2y_t-\mu y_u} u^{-\mu} \mathscr{C}^\pm(b^{y_t}\tau, 0). \tag{138}$$

Fixing b to $|\tau|^{-1/y_t}$ or $|h|^{-1/y_h}$ in (136)–(138), appropriately then gives[75]

$$\beta_{\mathrm{MFT}} = \frac{d - y_h}{y_t} + \frac{\kappa y_u}{y_t} = \frac{1}{n-2}, \tag{139}$$

$$\frac{1}{\delta_{\mathrm{MFT}}} = \frac{d - y_h}{y_h} + \frac{\lambda y_u}{y_h} = \frac{1}{n-1}, \tag{140}$$

$$\alpha_{\mathrm{MFT}} = \frac{2y_t - d}{y_t} - \frac{\mu y_u}{y_t} = \frac{n-4}{n-2}. \tag{141}$$

These now match Eqs.(97), (99) and (108) so the free energy sector in the RG formalism above d_{uc} is repaired. The amplitudes have also to be consistent and we find, for example, $B^- = \mathcal{M}^-(0^-)u^{-\kappa}$.

Following the principle of not repairing that which does not appear to need repairing, nothing has to be modified for the correlation sector.

We have now reached a point where RG appears to be successful in its treatment of critical properties above the upper critical dimension in the thermodynamic limit at least.

6.2. A problem with Finite-Size Scaling

There is, however a problem with FSS. Inserting the MFT exponents in the FSS prescription (10), one expects at the pseudo-critical point $t = 0$

$$c_L(t = 0, 0) \sim L^{\frac{\alpha_{\mathrm{MFT}}}{\nu_{\mathrm{MFT}}}} = L^{2\frac{n-4}{n-2}}, \tag{142}$$

$$\chi_L(t = 0, 0) \sim L^{\frac{\gamma_{\mathrm{MFT}}}{\nu_{\mathrm{MFT}}}} = L^2, \tag{143}$$

$$m_L(t = 0, 0) \sim L^{-\frac{\beta_{\mathrm{MFT}}}{\nu_{\mathrm{MFT}}}} = L^{-\frac{2}{n-2}}, \tag{144}$$

$$\xi_L(t = 0, 0) \sim L. \tag{145}$$

We call this *Landau FSS* because the exponents which appear in powers of L are all ratios of exponents from Landau theory. Here we have presented FSS at the pseudo-critical point $t = 0$ (see Eq.(12)) rather than the critical point $\tau = 0$ (see Eq.(9)). This is usually easier to work with numerically because simulations, like pseudo-critical points, necessitate finite size. In contrast, the critical temperature itself requires extrapolation to the limit of infinite volume. Usually (for example below the critical dimension), it makes little difference whether we implement FSS at $t = 0$ or $\tau = 0$. This is because both of them fall within the same scaling window - they exhibit the same FSS behaviour. This is the case if the rounding is not too narrow.

These standard FSS predictions, however, fail to match numerical simulations and exact results. In the early 1980's, Brézin[7] considered finite-size correlation length for the ϕ^4 model above $d_{\mathrm{uc}} = 4$ on a hypercubic systems with PBCs. His theoretical analysis predicted

$$\xi_L(\tau = 0, 0) \sim L^{d/4} \tag{146}$$

which contradicts Landau scaling (145). Then, in 1985, with Jean Zinn-Justin, Brézin presented a more complete analysis for the ϕ^4 model,[76] delivering

$$\chi_L(\tau = 0, 0) \sim L^{d/2}, \tag{147}$$

for the susceptibility behaviour, above d_{uc}. In the same year, Binder presented numerical results for the susceptibility of the Ising model in 5D with PBCs[20] (see also Ref. [77]). At the pseudo-critical point he obtained

$$\chi_L(t = 0, 0) \sim L^{5/2}, \quad \text{and} \quad |\tau_L| = |T_L/T_c - 1| \sim L^{-5/2}. \tag{148}$$

This is in agreement with the results of Brézin and Zinn-Justin, but disagrees with Landau FSS.

Thus, despite Fisher's DIVs rescuing the GFP RG formalism above d_{uc}, we arrive at a disagreement between Landau FSS and exact/numerical results. Even though we know that GFP values for the critical exponents do not match MFT, attempts to invoke them for standard FSS do not come to the rescue. While they deliver different FSS in the free-energy sector (namely, $c_\infty \sim L^{\alpha_G/\nu_G} = L^{4-d}$ and $m_\infty \sim L^{-\beta_G/\nu_G} = L^{-(d-2)/2}$), they necessarily deliver the same as Landau FSS for χ and ξ because the correlation sector is not (yet) perceived to be in danger. As we have seen, these are independent of n and do not agree with exact or numerical results.

Therefore, while RG may have been rescued above d_{uc}, by 1985 FSS was to be sacrificed. We next address an immediate attempt at resuscitation.

6.3. *Dangerous Irrelevant Variables for finite size: the "starred" Renormalization Group eigenvalues*

The consequences of the existence of DIVs for the finite-size scaling form of the free energy, and for hyperscaling, were investigated in 1985 in Ref. [78]. There, with Binder and Young, Michael Nauenberg and Vladimir Privman suggested to extend Fisher's ideas so that not only the observable thermodynamic functions but also the underlying free energy might be affected. We refer to the authors of this important paper as BNPY. In particular, BNPY suggested that [cf Eq.(13)]:

$$f_\infty(x,y,z) \overset{z\to 0}{=} z^{p_1} f_\infty(xz^{p_2}, yz^{p_3}, 0) \tag{149}$$

or

$$f_\infty(\tau,h,u) \overset{u\to 0}{=} b^{-d+p_1 y_u} u^{p_1} \mathscr{F}^{\pm}(b^{y_t+p_2 y_u}\tau u^{p_2}, b^{y_h+p_3 y_u} h u^{p_3}), \tag{150}$$

where p_1, p_2 and p_3 are constants feeding into κ, λ and μ used above. BNPY also introduced a "starred" notation for the renormalised RG eigenvalues:

$$d^* - d - p_1 y_u, \quad y_t^* = y_t + p_2 y_u, \quad y_h^* = y_h + p_3 y_u. \tag{151}$$

For the ϕ^n model of Eq.(93), differentiation of the DIV-affected free energy density wrt τ, and to h allows

$$\alpha_{\mathrm{MFT}} = \frac{2y_t^* - d}{y_t^*}, \ \beta_{\mathrm{MFT}} = \frac{d - y_h^*}{y_t^*}, \tag{152}$$

$$\gamma_{\mathrm{MFT}} = \frac{2y_h^* - d}{y_t^*}, \ \delta_{\mathrm{MFT}} = \frac{y_h^*}{d - y_h^*}. \tag{153}$$

Comparing with Fisher's (141) in Subsection 6.1, leads to

$$p_1 = 0, \quad p_2 = -\frac{2\kappa y_t}{d + 2\kappa y_u} = -\frac{2}{n}, \quad p_3 = -\frac{\kappa(2y_h - d)}{d + 2\kappa y_u} = -\frac{1}{n}, \qquad (154)$$

having used Eq.(132) for the general n case. Inserting these values in Eq.(151) gives

$$d^* = d, \quad y_t^* = y_t - \frac{2y_u}{n}, \quad y_h^* = y_h - \frac{y_u}{n}. \qquad (155)$$

In terms of d and n the starred scaling dimensions are

$$y_t^* = \frac{d(n-2)}{n}, \quad y_h^* = \frac{d(n-1)}{n}, \qquad (156)$$

and in terms of $\wp$ they read as

$$y_t^* = 2\wp, \quad y_h^* = \frac{d}{2} + \wp. \qquad (157)$$

By construction, inserting y_t^* for y_t and y_h^* for y_h in (20), (21), (22) and (23), delivers the correct Landau MFT critical exponents for α, β, δ and γ in (97), (99), and (103), (108). Again by construction, the remaining main critical exponents ν and η are left intact because the correlation sector is untouched in Eq.(150).

Having extended DIVs to the finite-size regime, DIVs also extend finite-size scaling for the susceptibility and magnetisation in terms of the starred exponents instead of the GFP ones. In particular, they find that

$$m_L \sim L^{y_h^* - d}$$

and

$$\chi_L \sim L^{2y_h^* - d}.$$

These are indeed correct as we shall see. However, in BNPY's formalism, they do not come from FSS in a form like Eq.(8).

BNPY made an insightful step analogous to Eq.(150) for the correlation length. They proposed that

$$\xi_\infty(\tau, h, u) \overset{u \to 0}{=} b^{1 + q_1 y_u} u^{q_1} \Xi(b^{y_t + q_2 y_u} \tau u^{q_2}, b^{y_h + q_3 y_u} h u^{q_3}), \qquad (158)$$

with the three parameters, q_1, q_2 and q_3 initially left open. Then they make the following crucial but erroneous assumption:

> "Since the finite-size correlation length ξ_L is bounded by L, we require $q_1 y_u < 0$. (...) if one adopts the plausible assumption that for $\tau = h = 0$, the correlation length increases up to the linear dimensions of the lattice, which implies that $q_1 = 0$."

We will see later that the value $q_1 = 0$ is not correct. It feeds into the exponent $ϙ$ in an essential way — the assumption that $q_1 = 0$ is equivalent to assuming $ϙ = 1$ as we shall see. BNPY did not pursue a discussion of q_2 or q_3, leaving open the possibility that they might differ from p_2 and p_3. We shall see below that, although this option has to be considered, this is not the case.

At this stage, we are still confronted with the unsatisfactory situation that, while all critical and finite-size scaling behaviour seems to be rescued, the FSS prescription itself fails above the upper critical dimension. In an attempt to replace that, a new length scale was introduced — to replace the (seemingly unbroken), correlation length, as the relevant scale which controls finite-size effects there. Several other length scales were invented by other authors and these are discussed in Volume IV.[21] We do not pursue these considerations here as we believe them to be redundant. Instead, we follow the outstanding work by Luijten and Blöte,[79,80] which uses RG to bring corrections to scaling into the DIV scenarios and to connect with much of the mathematical literature.

7. Scaling of the Fourier modes

Eq.(93) gives the free energy for the generic ϕ^n model, and its $n = 4$ realisation is

$$F_{\text{GLW}}[\phi(\mathbf{x})] = \int d^d x \left(\tfrac{1}{2} r \phi^2(\mathbf{x}) + \tfrac{1}{4} u \phi^4(\mathbf{x}) - h\phi(\mathbf{x}) + \tfrac{1}{2} |\nabla \phi|^2 \right). \tag{159}$$

If PBCs are used, this can be expressed in Fourier space as

$$F_{\text{GLW}}[\tilde{\phi}_{\mathbf{k}}] = \tfrac{1}{2} \sum_{\mathbf{k}} (r + |\mathbf{k}|^2)|\tilde{\phi}_{\mathbf{k}}|^2 - hL^{d/2}\tilde{\phi}_0$$

$$+ \tfrac{1}{4} u L^{-d} \sum_{\mathbf{k}_1, \mathbf{k}_2, \mathbf{k}_3} \tilde{\phi}_{\mathbf{k}_1} \tilde{\phi}_{\mathbf{k}_2} \tilde{\phi}_{\mathbf{k}_3} \tilde{\phi}_{-(\mathbf{k}_1 + \mathbf{k}_2 + \mathbf{k}_3)} \tag{160}$$

with

$$\phi(\mathbf{x}) = \sum_{\mathbf{k}} \tilde{\phi}_{\mathbf{k}} \psi_{\mathbf{k}} = \frac{1}{\sqrt{V}} \sum_{\mathbf{k}} \tilde{\phi}_{\mathbf{k}} e^{i\mathbf{k}\cdot\mathbf{x}}, \quad \mathbf{k} = \frac{2\pi}{L}\mathbf{n}, \ \mathbf{n} \in \mathbb{Z}^d. \tag{161}$$

The zero mode $\tilde{\phi}_0$ itself is worthy of being distinguished from other Fourier modes $\tilde{\phi}_{\mathbf{k}\neq 0}$ and we write

$$F_{\text{GLW}}[\tilde{\phi}_0, \tilde{\phi}_{\mathbf{k}\neq 0}] \simeq \tfrac{1}{2} \left(r + \frac{3u}{2L^d} \sum_{\mathbf{k}\neq 0} |\tilde{\phi}_{\mathbf{k}}|^2 \right) \tilde{\phi}_0^2 + \tfrac{1}{4} \frac{u}{L^d} \tilde{\phi}_0^4 - hL^{d/2}\tilde{\phi}_0$$

$$+ \tfrac{1}{2} \sum_{\mathbf{k}\neq 0} (r + |\mathbf{k}|^2)|\tilde{\phi}_{\mathbf{k}}|^2 + \cdots. \tag{162}$$

Here we have suppressed terms of higher order than Gaussian for the non-zero modes because only the zero mode contributes to the non-vanishing average order parameter below the critical temperature. The non-zero modes do not manifest non-vanishing average values even in the ordered phase (see Section 2). Therefore we have two terms corresponding to two types of Fourier mode. The DIV u enters alongside the temperature field r to couple with the quadratic term in the zero mode expansion but the non-zero modes do not suffer from this shift. This means that, while the zero mode $\tilde{\phi}_0$ is governed by the DIV-adjusted GFP, the non-zero modes $\tilde{\phi}_{\mathbf{k}\neq 0}$ are only controlled by the GFP itself.

We refer to the anomalous scaling emanating from DIVs as manifested in equations (146) or (147) as Q scaling. We discuss this in more detail in Section 9. In Ref. [81], Wittmann and Young analyzed non-zero modes and concluded *standard* FSS holds there. Thus one has for $\mathbf{k} \neq 0$,

$$\chi_{\mathbf{k}\neq 0} = L^d \langle |\tilde{\phi}_{\mathbf{k}\neq 0}|^2 \rangle_L \sim L^2. \tag{163}$$

This statement, however, hides another subtlety in the story of Q. Eq.(163) as *standard* FSS can either refer to Landau scaling, or to the GFP scaling (see Table 2) because both deliver an exponent 2. The standard picture is (or was) that *standard* FSS refers to the Landau picture. Indeed, in our early contributions to this topic,[6] although we referred to the $\chi_L \sim L^2$ behaviour as *Gaussian* scaling, we had Landau or MFT in mind. In a later publication, it became clear that it is Gaussian and not Landau scaling that is at play in the non-zero modes.[27] To disentangle them we looked at the FSS behaviour of the non-zero modes for magnetisation as well. There, Gaussian FSS predicts

$$m_{\mathbf{k}\neq 0} = \langle |\tilde{\phi}_{\mathbf{k}\neq 0}| \rangle_L \sim L^{-\frac{d-2}{2}}, \tag{164}$$

while Landau FSS would deliver L^{-1}. In Ref. [27], Eq.(164) was shown to indeed be correct. Nowadays we refer to the GFP scaling described in Section 5 as G scaling to distinguish it from the Q scaling. It is G-scaling (not Landau) that manifests as Eqs.(163) and (164), associated with non-zero modes. We keep the term standard FSS for that which comes from the Landau picture in Section 4 even though it has no direct physical manifestation.

8. Corrections to finite-size scaling and their dominance at short distances

Luijten and Blöte considered RG equations for the scaling fields in the $O(N)$ ϕ^4 model and we extended to the ϕ^n model (93) in Ref. [1]. The outcome is that the DIV contaminates not only the amplitudes but the manner in which the temperature and magnetic field enter the homogeneity assumption as well. The RG equations

$$\frac{dr}{d\ln b} = y_t r + pu, \quad \frac{du}{d\ln b} = y_u u, \quad \frac{dh}{d\ln b} = y_h h \qquad (165)$$

lead to free energy density taking the form

$$f_L(\tau, h, u) = L^{-d} \mathscr{F}^{\pm}[L^{y_t^*}(\tau u^{-2/n} - \tilde{p} u^{(n-2)/n} L^{y_u - y_t}), L^{y_h^*} h u^{-1/n}], \qquad (166)$$

where

$$\tilde{p} = -\frac{p}{(y_u - y_t)} \qquad (167)$$

and the temperature scaling field is

$$\tau = r + \tilde{p} u. \qquad (168)$$

From Eq.(166), two types of FSS for thermodynamic quantities emerge – leading and corrected.[1] It also governs the leading FSS of the pseudocritical temperature. Denoting the first argument by

$$X = L^{y_t^*}(\tau u^{-2/n} - \tilde{p} u^{(n-2)/n} L^{y_u - y_t}), \qquad (169)$$

we identify a fixed value of it, say X_0, as that for which a given thermodynamic variable in the form of a derivative — susceptibility, for example — peaks. This is the pseudocritical temperature τ_L, and we now have

$$\tau_L = X_0 u^{2/n} L^{-y_t^*} + \tilde{p} u L^{-(y_t - y_u)}. \qquad (170)$$

Because $y_t^* = y_t - \frac{2}{n} y_u \leq y_t - y_u$ above d_{uc}, the first term dominates for large enough L, so that the shift exponent is $\lambda = y_t^*$ instead of $\lambda = y_t = 1/\nu$ in Eq.(84). This recovers Eq.(148) for the 5D Ising model, for example.

In our opinion, the analytic inclusion of corrections to scaling in Eq.(166) is also a central result. The two-term structure was already proposed in BNPY for FBCs but not for PBCs. While they proposed that the free energy scales as Eq.(149) (with $b = L$) for PBCs,

> "for other boundary conditions, where the system has a surface, it is probably necessary to use both $\tau L^{y_v^*}$ and $\tau L^{1/\nu}$ for a complete asymptotic description".[78]

The second term in Eq.(170) when $n = 4$ corresponds to BNPY's proposal and extends it to other boundary conditions, including PBCs.

Luijten and Blöte also derived two different decay modes for the spin-spin correlation function on this basis. In Ref. [43], they differentiated the

---

[1] For a more elaborate discussion on the possible role of corrections, see Refs. [1,82].

finite-size free energy density wrt two *local* magnetic fields placed at positions 0 and $\mathbf{x}$, "assuming that the finite-size behaviour is identical to the $\mathbf{x}$ dependence of g." With this neat trick, they can account for DIVs through the free energy and incorporate them into the correlation function, without compromising the belief that the correlation length itself is anything other than linear in its dependency on finite size. As they say [we set the long-range interaction-decay exponent to 2 for the short-range model under consideration here],

> "If we do not take into account the dangerous irrelevant variable mechanism, we find $g \propto L^{2y_h - 2d} = L^{-(d-2)}$, just as we found before from $\eta = 0$. However, replacing y_h by y_h^* yields $g \propto L^{-d/2}$, in agreement with the L dependence of the magnetic susceptibility. This clarifies the difference between the two predictions: At short distances (large wave vectors), there is no "dangerous" dependence on u. Hence, the finite-size behaviour of the spin-spin correlation function will be given by $L^{-(d-2+\eta)}$. For $k = 0$, the coefficient of the ϕ^2 term vanishes and thus the $u\phi^4$ term is required which implies that y_h is replaced by y_h^* and g_L scales as $L^{2y_h^* - 2d}$."

Thus Luijten and Blöte arrived at two modes for the correlation function, the relative magnitudes of r and L determining which one applies. For short distances G modes dominate and for long distances Q modes reign. More explicitly, taking corrections into account, for Q modes they have[m]

$$g_Q(L/2) = L^{2y_h^* - 2d}[c_o + c_1 \hat{t} L^{y_t^*} + c_2 \hat{t}^2 L^{2y_t^*} + \ldots] \tag{171}$$

with $\hat{t} \equiv \tau + \mathrm{Const}\, L^{y_u - y_t}$. They have a counterpart formula for log corrections. For G modes, on the other hand, they have

$$g_G(L/2) = L^{2y_h - 2d}[c_o + c_1 \hat{t} L^{y_t} + c_2 \hat{t}^2 L^{2y_t} + \ldots]. \tag{172}$$

There are no logarithms at the upper critical dimension for G modes because G modes are not affected by dangerous irrelevancy.

Thus we have that to leading order, and above the critical dimension, $g_Q \sim g_G \times L^{-2y_u/n}$ and, since $y_u = -2(d - d_{\mathrm{uc}})/(d_{\mathrm{uc}} - 2)$ is negative there, g_Q decays slower than g_G. To determine if the two-decay-mode structure is visible in finite systems they simulated the long-range Ising model in one dimension and found that, indeed it is — provided the system size is large enough to leave enough space to see it. Since this observation has been overlooked in recent literature[83] we reproduce it here (Fig. 1).

---

[m]Of course Luijten and Blöte don't use the notation g_G for G-modes and g_Q for Q-mode, so we are somewhat preemptive in introducing our notation at this point.

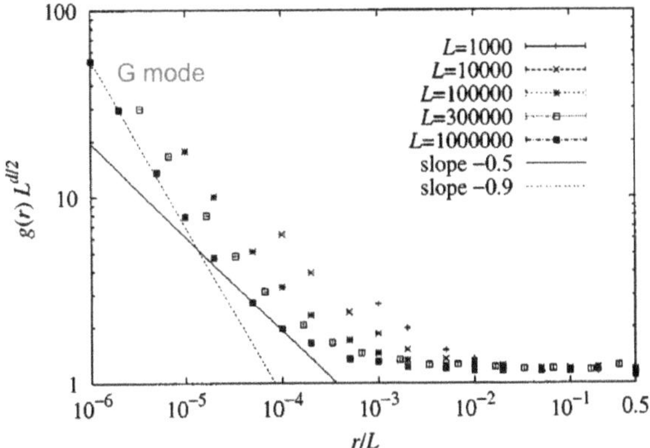

Figure 1.   The spin-spin correlation function versus r/L in the 1D model with $\sigma = 0.1$. Figure from Ref. [43].

In terms of critical exponents, the more traditional way of presenting the correlation function is through the scaling dimension as in Eq.(14) or the anomalous dimension as in Eq.(24). In these terms, the Q-mode decay at $\tau = 0$ is

$$g(\mathbf{x}) \sim \frac{1}{|\mathbf{x}|^{2d-2y_h^*}} = \frac{1}{|\mathbf{x}|^{2x_\phi^*}} = \frac{1}{|\mathbf{x}|^{d-2+\eta^*}} = \frac{1}{|\mathbf{x}|^{d-2\varrho}} = \frac{1}{|\mathbf{x}|^{2d/n}}, \qquad (173)$$

with $\eta^* = d + 2 - 2y_h^* = 2 - 2\varrho$, while the G-mode decay is

$$g(\mathbf{x}) \sim \frac{1}{|\mathbf{x}|^{2d-2y_h}} = \frac{1}{|\mathbf{x}|^{2x_\phi}} = \frac{1}{|\mathbf{x}|^{d-2+\eta}} = \frac{1}{|\mathbf{x}|^{d-2}}. \qquad (174)$$

Quoting Luijten and Blöte again,

> "The fact that the L dependence of $g(L/2)$ is determined by the
> $k = 0$ mode raises the question of whether one can also observe
> the power-law decay described by η [$= 0$] in finite systems."

To this end, they sampled the correlation decay as a function of $\mathbf{x}$ far into the high-dimensional regime. They found that very large lattice sizes are required to leave enough space to observe the short-distance, $\eta = 0$, G mode decay at the start (for small $\mathbf{x}$ relative to large L). As stated by Luijten and Blöte, for $\mathbf{x}$ of the order of the system size,

> "the correlation function levels off. This is the mean-field con-
> tribution to the correlation function which dominates in the
> spatial integral yielding the magnetic susceptibility."

In an attempt to fit it into the Q-scheme yet another approach was introduced in Ref. [28]. The standard derivation of Fisher's fluctuation-response relation (30) involves integrating the correlation function (25) over the full infinite-size correlation volume ξ_∞^d to obtain the susceptibility close to the critical point. The same scaling relation ensues for finite volume provided the correlation length ξ_L is bounded by the length. However, we now know this is not the case. According to this view, the integral limits should depend upon whether one integrates over a scale of length L and dimension d or correlation length ξ_L and dimension $d_{\rm uc}$. This delivers two different anomalous dimensions — $\eta = 0$ on the scale of ξ_L and $\eta_Q = 2 - \digamma\gamma/\nu$ on the scale of L. Thus the standard version of Fisher's scaling relation (30) holds in high dimensions if length-scale is taken as correlation length. A corresponding counterpart to the fluctuation-dissipation relation (25) ensues and is

$$\eta_Q = \digamma\eta + 2(1 - \digamma). \tag{175}$$

The issue of correlation decay in high dimensions had previously been addressed by John Nagle and Jill Bonner in 1970.[84] In a numerical study, they found a negative value for the anomalous dimension of a one-dimensional chain with long-range interactions which took it above the upper criticality. They termed this $\tilde{\eta}$. George Baker and Geoffrey Golner attempted to provide an explanation a few years later.[85] They suggested a difference between "short long-range order" and "long long-range order" with only the former controlled by the standard anomalous dimension. The negative anomalous dimensions of Nagle, Bonner, Luijten and Blöte come from measuring distance on a scale of ξ_L. Thus the η^* of Luijten and Blöte is the same as the $\tilde{\eta}$ of Nagle and Bonner.

Finite-size effects in the high-dimensional Ising model were investigated in an outstanding PhD thesis by Vassilios Papathanakos (under the supervision of Michael Aizenman) in 2006.[86]

> "These results are consonant with recent findings in the theories
> of percolation and loop-erased random walks, which contribute
> to an emerging picture of multi-scale criticality."

The effect of boundary conditions was investigated using a random-geometric approach coupled with a rigorous study of the Ising model on a high-dimensional box of volume L^d. Papathanakos found that

> "the short-range behaviour of the model is essentially unaffected
> by the choice of boundary conditions, and can be understood
> in terms of non-intersecting, independent simple random walks;
> this is reflected in the behaviour of bulk quantities, at temper-
> atures where the correlation length is small enough. However,
> as the critical temperature is approached, the probability for

the intersection of the paths in the case of periodic boundary conditions becomes significant: the reentrance temperature sets the scale for the critical susceptibility, which is found to be at least of order $[\xi_L(T_c) \gtrsim L^{d/2}]$ in contrast to the case of bulk boundary conditions, where $[\xi_L(T_c) \lesssim L^{d/2}]$.

Moreover, the periodic two-point function at criticality can be essentially decomposed as the sum of the infinite-lattice two-point function, and a mean-field plateau, which dominates in the bulk. Furthermore, making the assumption that the susceptibility is not larger than $[O(L^{d/2})]$, we can also show that the renormalized coupling constant $g_L(T_c)$ of the system with periodic boundary conditions has a non-vanishing scaling limit, unlike the case of bulk boundary conditions, where the scaling limit is Gaussian. This assumption also implies that the value of the plateau at criticality is fixed at $[g_L = O(L^{-1/2})]$. These findings are in essential agreement with recent numerical results present in the literature, and forward a more clear and intuitive explanation than theoretical estimates based on renormalization group arguments."

The main results for the Ising model are that the FSS for the susceptibility with FBCs is $\chi_L(T_c) \sim L^2$ while that for PBCs is $\chi_L(T_c) \sim L^{d/2}$. Also for PBCs, the correlation function has reentrant behaviour vis

$$
g_L(T_c, \mathbf{x}) \sim
\begin{cases}
\frac{C_1}{|\mathbf{x}|^{d-2}} & \text{if } |\mathbf{x}| \leq C_3 L^{\frac{d}{2(d-2)}} \\[2ex]
\frac{C_2}{|\mathbf{x}|^{1/2}} & \text{if } |\mathbf{x}| \geq C_3 L^{\frac{d}{2(d-2)}}.
\end{cases}
\tag{176}
$$

This brings us to the close of the century. But this by no means closes the story. In 2000 Binder and Luijten say:[87]

"Another issue where there has been a longstanding controversy between theory and simulation is the question of finite size scaling at dimensionalities d above the marginal dimension d^* where mean field theory becomes valid [22–30]. For systems with short-range forces, $d^* = 4$, so in $d = 5$ all critical exponents (including corrections to scaling) are known, and hence finite-size scaling methods can be exposed to a stringent test. However, it will be shown that the comparison between theory [29] and simulation [30] is still disappointing!"

They conclude with:[87]

"In this work, two simple aspects of the bulk critical behaviour of ferromagnetic Ising models were discussed, and it was shown that both problems (crossover from one universality class to another, and finite-size scaling above the marginal dimension) still are incompletely understood."

9. Extension of dangerous irrelevancy in the correlation sector: the use of ϙ, the rescue of hyperscaling and finite-size scaling

Forty years ago Fisher rescued the RG framework for the free-energy sector in the thermodynamic limit (Sec. 6.1) above the upper critical dimension, via his introduction of DIVs which were then extended to finite-size scaling corrections (Sec. 8). However, problems still remain and, following Ref. [1], we further build on Fisher's legacy to preserve the RG framework, extending DIVs to the correlation sector. In doing so, we closely follow Luijten and Blöte's approach that we proposed to generalize in Ref. [27].

We return firstly to Section 6.3 and BNPY's extension of the DIV mechanism to the free energy itself. As stated, they assumed that the correlation length could obey a similar homogeneity law (158) but insisted that $q_1 = 0$ and left q_2 and q_3 open. For the free energy in Eq.(150), BNPY gave three reasons why $p_1 = 0$ and hence $d^* = d$. In a series of papers, *e.g.* in Ref. [8], we advocated that the correlation length should be allowed to exceed the length — as Brézin had shown for $d = 4$ and general n. In particular, we showed that

$$\xi_L \sim L^{\text{ϙ}}, \tag{177}$$

with

$$\text{ϙ} = d/d_{\text{uc}}. \tag{178}$$

Therefore we write

$$\xi_L(t, h, u) = b^{\text{ϙ}} \Xi^{\pm}(b^{y_t^+} t u^{q_2}, b^{y_h^+} h u^{q_3}), \tag{179}$$

where

$$\text{ϙ} = 1 + q_1 y_y, \quad y_t^+ = y_t + q_2 y_u, \quad y_h^+ = y_h + q_3 y_u. \tag{180}$$

The value for ϙ in Eq.(178) determines $q_1 = -1/n$. In the thermal sector, matching Eq.(179) with the value $\nu_{\text{MFT}} = 1/2$ in Eq.(116) determines $q_2 = -2/n$, identical to p_2. Likewise, matching Eq.(179) with the value $\nu_{\text{c MFT}} = (n-2)/2(n-1)$ in Eq.(116) determines $q_3 = -1/n$, identical to p_3. Note that, while the correct correlation-length critical exponent ν_{MFT} is identical to that coming from the GFP $\nu_{\text{GFP}} = 1/y_t = 1/2$ in Eq.(25), their counterparts in the magnetic sector do not coincide and $\nu_{\text{c MFT}}$ differs from $\nu_{\text{c GFP}} = 1/y_h$ from Eq.(26). So, technically although $\nu_{\text{MFT}} = 1/y_t$, this is by coincidence only and the correct expression is $\nu_{\text{MFT}} = \text{ϙ}/y_t^+ = \text{ϙ}/y_t^*$. Note also that because $q_2 = p_2$ and $q_3 = p_3$, one has $y_t^+ = y_t^*$ and $y_h^+ = y_h^*$.

We henceforth only use the starred scaling dimensions. *I.e.*, to extend Eqs.(152) and (153) to the correlation sector, we write

$$\nu_{\mathrm{MFT}} = \frac{\digamma}{y_t^*}, \qquad \nu_{\mathrm{c\,MFT}} = \frac{\digamma}{y_h^*}. \tag{181}$$

Equations (155) and (156) can now be written above d_{uc} in a very consistent manner as

$$d^* = \digamma d_{\mathrm{uc}}, \quad y_t^*(d) = \digamma y_t(d_{\mathrm{uc}}), \quad y_h^*(d) = \digamma y_h(d_{\mathrm{uc}}). \tag{182}$$

Likewise, we believe that there is no reason to invoke a mechanism by which the rescaling of the temperature and the magnetic field would differ for different physical quantities. Therefore, the result that $y_t^+ = y_t^*$ and $y_h^+ = y_h^*$ is satisfactory. A second extension proposed is to use the correct (starred) scaling dimension ($2x_\phi^*$ in fact) of the matter field to describe the dimension of the correlation function. From these considerations, we are led to the following scaling hypotheses (Eq.(166) is rewritten for the sake of clarity),

$$f_L(\tau, h, u) = b^{-d}\mathscr{F}^{\pm}(X, Y, bL^{-1}), \tag{183}$$

$$g_L(\mathbf{x}, \tau, h, u) = b^{-2x_\phi^*}\mathscr{G}^{\pm}(b^{-1}\mathbf{x}, X, Y, bL^{-1}), \tag{184}$$

$$\xi_L(\tau, h, u) = b^{\digamma}\Xi^{\pm}(X, Y, bL^{-1}). \tag{185}$$

These are expressed in terms of the two rescaled variables

$$X = b^{y_t^*}(\tau u^{-2/n} - \tilde{p}u^{(n-2)/n}b^{y_u - y_t}), \tag{186}$$

$$Y = b^{y_h^*}hu^{-1/n}, \tag{187}$$

and with the scaling dimensions

$$y_t^* = d\left(1 - \frac{2}{n}\right), \ y_h^* = d\left(1 - \frac{1}{n}\right), \ x_\phi^* = \frac{d}{n}, \ \digamma = \frac{d}{2}\left(1 - \frac{2}{n}\right) \tag{188}$$

having used (155) with (121), (122), (123) and extended x_ϕ^* from Ref. [27] for the $n = 4$ case, and where $\tilde{p}$ is given in Eq.(167). Obviously these scaling forms apply to the Q-sector only (Fourier Q-modes). The G sector is unaffected by DIVs and the finite-size counterparts of Eqs.(13), (14) and (15) apply there. Following a suggestion by Michael Fisher, a new exponent $\digamma$ was introduced in Ref. [8] for the Q sector. We like the use of this exponent, since it is very easy to translate equations from one universality class to another in terms of d/d_{uc}, and also to generalize from the GFP values (which we recover reverting $\digamma$ to 1), so one can also rewrite the exponents (188) in the form

$$y_t^* = 2\digamma, \ y_h^* = \frac{d}{2} + \digamma, \ x_\phi^* = \frac{d}{2} - \digamma, \ \digamma = \frac{d}{d_{\mathrm{uc}}}. \tag{189}$$

An interesting use of φ is in a new form of the hyperscaling relation. Setting $b = |\tau|^{-1/y_t}$ in (185) delivers $\xi_\infty \sim |\tau|^{-\varphi/y_t^*}$, hence $y_t^* = \varphi/\nu$, then in (183) we get $f_\infty \sim |t|^{d/y_t^*} \sim |\tau|^{2-\alpha}$ which leads to

$$\alpha = 2 - \frac{\nu d}{\varphi}. \tag{190}$$

This repairs[n] the hyperscaling relation above d_{uc}, and it holds also below d_{uc} where $\varphi = 1$. (Obviously $\varphi = 1$ for the non-zero or orthogonal Fourier modes in high dimensions too.)

The new exponent already enabled predictions to be more naturally expressed for models such as the nearest-neighbour Ising model, percolation above its critical dimension $d_{\mathrm{uc}} = 6$ and for Long-Range Ising Models (LRIM' with various dimensions above $d_{\mathrm{uc}}(\sigma)$ (with periodic boundary conditions).[60,89] The extension to general values of n is obvious and we collect the predictions for arbitrary $d > d_{\mathrm{uc}}$ for quantities which have been discussed above:

$$m_L \sim L^{-(\frac{d}{2}-\varphi)}, \ \chi_L \sim L^{2\varphi}, \ e_L \sim L^{-(d-2\varphi)}, \ c_L \sim L^{4\varphi-d}, \tag{191}$$

$$\xi_L \sim L^\varphi, \ g_L(X_0) \sim L^{-(d-2\varphi)}, \tag{192}$$

$$t_L \sim L^{-2\varphi}, \ \Delta\beta_L \sim L^{-2\varphi}, \ |h^L| \sim L^{-(\frac{d}{2}+\varphi)}, \tag{193}$$

$$h_L^{\mathrm{LY}} \sim L^{-(\frac{d}{2}+\varphi)}, \ t_L^{\mathrm{F}} \sim L^{-2\varphi}. \tag{194}$$

Note that all of the above scaling formulas are readily obtained by replacing the standard FSS prescription Eq.(10) by the prescription

$$Q_\infty(t,0) \sim |t|^\rho \longrightarrow Q_L(t=0,0) \sim L^{-\varphi\rho/\nu}. \tag{195}$$

This is what we call Q-scaling and it holds at the pseudocritical point as we shall see shortly. With Q-scaling to hand, FSS holds above the upper critical dimension.

The different approaches that we have discussed are collected and compared in Table 2 for the ϕ^4 model with periodic boundary conditions. In this table, the first column lists known results in the thermodynamic limit or for finite-size scaling. The remaining columns correspond to the different approaches that we have described so far, with the symbol $\sqrt{}$ to denote an agreement and, in cases of disagreement, the prediction made by the

---

[n]Since Josephson's inequality $\nu d \geq 2 - \alpha$ was introduced in 1967,[88] literature, including textbooks, on statistical physics, lattice field theory, *etc.* refer to hyperscaling as "failing" above the upper critical dimension. This statement should no longer be used in statistical physics — hyperscaling does not fail because the RG does not fail above the upper critical dimension. Moreover, the hyperscaling relation should rather be rewritten properly as (190) and not as (27).

Table 2. Summary of the evolution of the scaling picture above the upper critical dimension for the ϕ^4 model. The first column presents the correct results (FSS predictions are for a system with periodic boundary conditions). In the other columns, we give the (incorrect) results predicted when they are different. A question mark means that the quantity hasn't been considered in the corresponding scenario.

The correct results	Landau scaling[1]	GFP[2]		Fisher DIV[3]	BNPY[4]	ϙ[5]
thermodynamic limit:						
$c_\infty(\tau,0) \sim |\tau|^0$	$\checkmark$	$\alpha_G = 2 - \frac{d}{2}$	$\checkmark$	$\checkmark$	$\checkmark$
$m_\infty(\tau,0) \sim |\tau|^{1/2}$	$\checkmark$	$\beta_G = \frac{d-2}{4}$	$\checkmark$	$\checkmark$	$\checkmark$
$\chi_\infty(\tau,0) \sim |\tau|^{-1}$	$\checkmark$	$\checkmark$		$\checkmark$	$\checkmark$	$\checkmark$
$m_\infty(0,h) \sim |h|^{1/3}$	$\checkmark$	$\delta_G = \frac{d+2}{d-2}$	$\checkmark$	$\checkmark$	$\checkmark$
$\xi_\infty(\tau,0) \sim |\tau|^{-1/2}$	$\checkmark$	$\checkmark$		$\checkmark$	?	$\checkmark$
$g_\infty(\mathbf{x},0,0) \sim |\mathbf{x}|^{-(d-2)}$	$\checkmark$	$\checkmark$		$\checkmark$	?	$\checkmark$
FSS:						
$\xi_L(\tau=0,0) \sim L^{d/4}$	L	L		L	L	$\checkmark$
$\chi_L(\tau=0,0) \sim L^{d/2}$	L^2	L^2		L^2	$\checkmark$	$\checkmark$
$\tau_L(\tau=0,0) \sim L^{-d/2}$	$L^{-1/2}$	$L^{-1/2}$		$L^{-1/2}$	$\checkmark$	$\checkmark$
$m_L(\tau=0,0) \sim L^{-d/4}$	L^{-1}	$L^{1-d/2}$		L^{-1}	$\checkmark$	$\checkmark$
$g_L(L/2,\tau=0,0) \sim L^{-d/2}$	$L^{-(d-2)}$	$L^{-(d-2)}$		$L^{-(d-2)}$	?	$\checkmark$
$h_L^{\mathrm{LY}}(\tau=0) \sim L^{-3d/4}$	L^{-3}	$L^{-(d+2)/2}$				$\checkmark$
$\tau_L^{\mathrm{F}}(\tau=0) \sim L^{-d/2}$	L^{-2}	L^{-2}				$\checkmark$

[1]FSS with Landau exponents,
[2]Predictions from the RG eigenvalues at the Gaussian Fixed Point,
[3]Corrections made by the scenario of Fisher,
[4]Most of the results presented in this column correspond to BNPY's version of the scenario of Fisher and are checked in Ref. [20],
[5]Q scaling of the results presented in the last column are checked *e.g.* in Ref. [6].

(incorrect) theory considered is given explicitly. The last column is for Q scaling.

10. The case of Free Boundary Conditions

10.1. *A mismatch between T_L and T_c in free boundary conditions*

The case of free boundary conditions is more problematic. First the easy part: the study of FSS properties of physical quantities evaluated *at the pseudo-critical point* for FBCs agrees with what we said previously for systems with PBCs. What is happening right at the critical point on the other hand is still subject to debate.

Lundow and Markström have studied this problem intensively for the Ising model in the 5D case, for very large systems (up to $L = 160$ in Ref. [90]). They obtained, for the magnetisation and the susceptibility, the leading behaviours and corrections to scaling of the form:

$$\chi_L(T_c) = 0.817\, L^2 + 0.083\, L, \tag{196}$$

$$m_L(T_c) = 0.230\, L^{-3/2} + 1.101\, L^{-5/2} - 1.63\, L^{-7/2}. \tag{197}$$

There, the leading and correction exponents are fixed and the coefficients are free, and the authors concluded in favour of a standard FSS behaviour of the susceptibility at the critical temperature $\chi_L(T_c) \sim L^2$. The same conclusion, albeit with much lower accuracy, was reached in Ref. [6]. There, by *standard* it was referred to FSS with Landau exponents, $\chi_L \sim L^{\gamma_{\rm MFT}/\nu_{\rm MFT}}$, a conclusion that we will see is not correct, although L^2 is (accidentally) correct.

The case of the magnetisation in (197) is a bit more subtle, because the leading exponent $-\frac{3}{2}$ is not the standard Landau FSS (the ratio of MFT exponents would be $-\beta_{\rm MFT}/\nu_{\rm MFT} = -1$ instead).

In order to clarify the situation, Wittmann and Young,[81,91] and then Flores-Sola *et al.*[27] considered the behaviour of the Fourier modes in a finite system with free boundary conditions. Following Ref. [92], a sine-expansion of the scalar field in the ϕ^4 action is performed with the boundary conditions $\phi(\mathbf{x}) = 0$ at the free surfaces.

Wittmann and Young confirmed the scaling at T_c for the total susceptibility (for FBC). They found that the single-mode susceptibility also obeys a standard FSS behaviour, $\chi_{\mathbf{k}} \sim L^2$ for the modes which will not acquire a nonzero magnetisation, namely with the smallest wave-vector with an even wave number n_α in the wave expansion ($k_\alpha = n_\alpha \pi/(L+1)$, $n_\alpha = 1, 2, \ldots, L$, $\alpha = 1, \ldots, d$). Wittmann and Young proposed to use $t = T - T_L$, with $t = \tau + {\rm const} \times L^{-\lambda}$, as the temperature-like scaling variable and the starred RG dimensions in the scaling hypothesis for the susceptibility. They obtain then $\chi_L(t) = L^{2y_h^* - d}\mathscr{X}(L^{y_t^*} t)$ in zero magnetic field and the compatibility with the bulk behaviour in the thermodynamic limit $\chi_\infty(\tau) \sim |\tau|^{-\gamma_{\rm MFT}}$ is recovered by the demand that the asymptotic regime obeys $\mathscr{X}(x) \sim x^{-\gamma_{\rm MFT}}$ (with $\gamma_{\rm MFT} = 1$ here) for $x \to \infty$. Thid leads to $\chi_L(t) \sim L^{2y_h^* - d + y_t^*}|t|^{-1}$. When $\lambda = d/2$, as is the case in PBC, this leads at criticality to the correct result $\chi_L(T_c) = L^{d/2}$. On the other hand if $\lambda = 2$ (FBC), this amounts to $\chi_L(T_c) = L^2$, hence the scenario proposed is very compelling.

The discussion concerning standard FSS with Landau exponents, or Gaussian FP exponents is nevertheless not settled by the study of the susceptibility alone and here we will present additional unpublished numerical results[26] for low-dimensional Ising models with long-range interactions above their respective upper critical dimensions.

10.2. *The three possible scenarios*

As we have already noted several times, one cannot disentangle, with the susceptibility alone, the predictions of standard (or Landau) FSS from those of the Gaussian Fixed point FSS. The primary aim of Ref. [27] was to test the discriminating case of the magnetisation. The argument of Wittmann and Young for the magnetisation would indeed lead to $m_L(T_c) = L^{-\lambda/2}$, hence $m_L(T_c) = L^{-d/4}$ in PBC and $m_L(T_c) = L^{-1}$ in FBC. The argument is thus equivalent to Landau scaling, because for FBC, $\lambda = 1/\nu_{\mathrm{MFT}}$.

So we are led to the point where one essentially faces three distinct hypotheses[o] among which one has to discriminate (see Table 3).

Table 3. The three scenarios for the FSS of the susceptibility and the magnetisation at T_c. There, $ϙ = d/d_{\mathrm{uc}}$, d_{uc} and the MFT exponents take their usual values and extension to different universality classes is obvious.

Hypothesis	Scaling of $\chi_L(T_c)$	Scaling of $m_L(T_c)$
1. Landau (or standard) scaling	$L^{\frac{\gamma_{\mathrm{MFT}}}{\nu_{\mathrm{MFT}}}}$	$L^{-\frac{\beta_{\mathrm{MFT}}}{\nu_{\mathrm{MFT}}}}$
2. Q scaling	$L^{2ϙ}$	$L^{ϙ-d/2}$
3. G scaling	L^2	$L^{1-d/2}$

In the case of the ϕ^4 universality class, the options are the following:

— prediction 1 (Landau scaling) leads to $\chi_L(T_c) \sim L^2$ and $m_L(T_c) \sim L^{-1}$,
— prediction 2 (Q scaling) to $\chi_L(T_c) \sim L^{d/2}$ and $m_L(T_c) \sim L^{-d/4}$,
— prediction 3 (G scaling) to $\chi_L(T_c) \sim L^2$ and $m_L(T_c) \sim L^{-(d-2)/2}$.

Option 2 is clearly ruled out at T_c by the results of Lundow and Markström in (196) (as well as by the results of Wittmann and Young). Options 1 and 2 are compatible with the results measured for the susceptibility. For the 5D model, it is easy to discriminate, with the magnetisation, between option 1 (which predicts the value -1), option 2 (prediction -1.25) and option 3 (prediction -1.5). Eq.(197) is in favour of option 3.

---

[o]In Ref. [1], a fourth scenario has been considered in which corrections to scaling could be dominant at small sizes, but this hypothesis has not been confirmed by the numerics. Therefore we do not discuss this fourth case further here.

    R. Kenna and B. Berche

10.3. *Towards a Gaussian Fixed Point scaling at T_c*

In Ref. [27] the magnetisation of the LRIM was investigated, tuning the parameter σ of the interaction decay to control the upper critical dimension $d/(2\sigma)$. In the thesis of Emilio Flores-Sola[26] results are reported for various values of σ and of d. Some of these are collected in Tables 4, 5 and 6 for the magnetisation, the susceptibility, correlation length and correlation function at T_c and at T_L. In the case of the magnetisation (Table 4), the expectation from GFP scaling is an exponent $-\beta_G/\nu_G$ with $\beta_G = (d-\sigma)/(2\sigma)$ and $\nu_G = 1/\sigma$, while Q scaling predicts an exponent $-\varsigma\beta_{\mathrm{MFT}}/\nu_{\mathrm{MFT}} = -\varsigma$.

Other quantities (temperature shift, correlation length, correlation function) are also reported in Ref. [26] (and partially collected here) and, although perfectible, all numerical results at T_c support the option 3 above (and also confirm Q scaling, *i.e.* option 2, at T_L). A recent work[93] has

Table 4. FSS of the magnetisation (upper panel) and the susceptibility (lower panel) for the LRIM in $d = 1, 2$ with FBCs at T_c and at T_L. The results at the critical temperature support G scaling ($\frac{d-\sigma}{2}$, resp. σ) while those at the pseudo-critical temperature support Q scaling ($d/4$, resp. $d/2$).

		$m_L(T_c)$	$\frac{d-\sigma}{2}$	$m_L(T_L)$	$\frac{\varsigma\beta_{\mathrm{MFT}}}{\nu_{\mathrm{MFT}}} = \frac{d}{4}$
$d = 1$	$\sigma = 0.1$	$L^{-0.450(4)}$	0.45	$L^{-0.233(4)}$	0.25
	$\sigma = 0.2$	$L^{-0.401(3)}$	0.40	$L^{-0.230(4)}$	0.25
$d = 2$	$\sigma = 0.1$	$L^{-0.949(1)}$	0.95	$L^{-0.501(2)}$	0.50
	$\sigma = 0.2$	$L^{-0.897(1)}$	0.90	$L^{-0.494(1)}$	0.50
		$\chi_L(T_c)$	σ	$\chi_L(T_L)$	$\frac{\varsigma\gamma_{\mathrm{MFT}}}{\nu_{\mathrm{MFT}}} = \frac{d}{2}$
$d = 1$	$\sigma = 0.1$	$L^{0.099(1)}$	0.1	$L^{0.522(3)}$	0.5
	$\sigma = 0.2$	$L^{0.200(1)}$	0.2	$L^{0.525(5)}$	0.5
$d = 2$	$\sigma = 0.1$	$L^{0.094(2)}$	0.1	$L^{0.985(2)}$	1.
	$\sigma = 0.2$	$L^{0.198(2)}$	0.2	$L^{0.994(2)}$	1.

Table 5. FSS of the correlation length for the LRIM in $d = 1$, 2 with FBCs at T_c and at T_L. The results at the critical temperature support G scaling (1) while those at the pseudo-critical temperature are more in favour of Q scaling ($d/(2\sigma)$).

		$\xi_L(T_c)$	1	$\xi_L(T_L)$	$\varsigma = \frac{d}{2\sigma}$
$d = 1$	$\sigma = 0.1$	$L^{1.01(3)}$	1	$L^{4.03(7)}$	5
	$\sigma = 0.2$	$L^{1.03(2)}$	1	$L^{2.21(4)}$	2.5
$d = 2$	$\sigma = 0.1$	$L^{1.07(7)}$	1	$L^{7.48(4)}$	10
	$\sigma = 0.2$	$L^{0.95(6)}$	1	$L^{3.97(4)}$	5

Table 6. FSS of the correlation function at $\mathbf{x}| = L/2$ for the LRIM in $d = 1$, 2 with FBCs at T_c and at T_L. The results at the critical temperature are more in favour of G scaling $(d - \sigma)$ while those at the pseudo-critical temperature support Q scaling $(d/2)$.

		$g_L(T_c, \frac{L}{2})$	$d - 2 + \eta_G$	$g_L(T_L, \frac{L}{2})$	$d - 2 + \eta_\varrho$
$d = 1$	$\sigma = 0.1$	$L^{-0.86(6)}$	0.9	$L^{-0.487(5)}$	0.5
	$\sigma = 0.2$	$L^{-0.83(6)}$	0.8	$L^{-0.483(6)}$	0.5
$d = 2$	$\sigma = 0.1$	$L^{-2.07(9)}$	1.9	$L^{-0.954(3)}$	1
	$\sigma = 0.2$	$L^{-1.70(9)}$	1.8	$L^{-0.974(3)}$	1

revisited FSS in the 5D Ising model (with short-range interactions) with FBCs and also confirms the present results.

The picture now is the following for FBCs: *Either* we study physical quantities which are related to the Q modes (*i.e.* modes that have a non-zero projection on the average magnetisation) or those related to the G modes (modes that do not contribute to the average magnetisation). In the first case, which is expected for FBC at T_L, the DIV has to be taken into account and Q scaling rules (option 2 in Table 3). In the second case, which holds for FBC at T_c, u is not dangerous and the physics is controlled by the GFP (option 3 in Table 3). The exponents there are Gaussian exponents collected in Table 1 *and not Landau exponents*.

11. Conclusion

We have reached a point where we believe that ϱ and $\hat{\varrho}$ are necessary ingredients of FSS at and above the upper critical dimension. They allow for a natural, and simple, extension of the scaling hypothesis that governs the behaviour of thermodynamic averages and correlations in systems having a non-zero finite-size order parameter. This is the case at the pseudo-critical points T_L, and also at T_c for systems with PBCs. There, the scaling hypothesis for the singular part of the free energy density, for the correlation length and for the order parameter correlation function can be written as follows

$$d > d_{\mathrm{uc}} \tag{198}$$

$$f(\tau, h, L^{-1}) = b^{-d} f(b^{2\varrho}\tau, b^{d/2+\varrho}h, bL^{-1}),$$
$$\xi(\tau, h, L^{-1}) = b^{\varrho}\xi(b^{2\varrho}\tau, b^{d/2+\varrho}h, bL^{-1}),$$
$$g(\tau, h, L^{-1}, \mathbf{x}) = b^{-(d-2\varrho)}g(b^{2\varrho}\tau, b^{d/2+\varrho}h, bL^{-1}, b^{-1}\mathbf{x}),$$

$$d = d_{\mathrm{uc}} \tag{199}$$

$$f_{\mathrm{uc}}(\tau, h, L^{-1}) = b^{-d_{\mathrm{uc}}} f(b^2 (\ln b)^{\hat{y}_t} \tau, b^{d_{\mathrm{uc}}/2+1} (\ln b)^{\hat{y}_h} h, bL^{-1}),$$

$$\xi_{\mathrm{uc}}(\tau, h, L^{-1}) = b \, (\ln b)^{\hat{\digamma}} \xi(b^2 (\ln b)^{\hat{y}_t} \tau, b^{d_{\mathrm{uc}}/2+1} (\ln b)^{\hat{y}_h} h, bL^{-1}),$$

$$g_{\mathrm{uc}}(\tau, h, L^{-1}, \mathbf{x}) = b^{-(d-2)} g(b^2 (\ln b)^{\hat{y}_t} \tau, b^{d_{\mathrm{uc}}/2+1} (\ln b)^{\hat{y}_h} h, bL^{-1}, b^{-1}\mathbf{x})$$

and the scaling of other quantities follows from appropriate derivatives wrt the scaling fields.

Hyperscaling relation among critical exponents and "hatted" critical exponents of the logarithmic corrections take the form

$$\alpha = 2 - \frac{\nu d}{\digamma}, \tag{200}$$

$$\hat{\nu} = \hat{\digamma} - \frac{\nu \hat{\alpha}}{2 - \alpha}. \tag{201}$$

They are valid above, below, and at d_{uc} for the first one (with $\digamma = 1$ below and $\digamma = d/d_{\mathrm{uc}}$ above d_{uc}), and the second one is valid at d_{uc}.

In FBCs, a special situation arises due to the rounding which does not scale like the shift of the pseudo-critical temperature, and, as a consequence, the pseudo-critical temperature is somehow pushed far below the critical temperature. It follows that at the critical temperature, there is no room for the order parameter to develop in the free energy mode expansion (it develops at the pseudo-critical point). The non-zero order parameter (this would be the zero-mode in the PBC case), which is responsible for the contamination of the scaling by the dangerous irrelevant variable, has no influence there. It follows that the correct homogeneity assumption that describes the singular behaviour in the very vicinity of the critical temperature $\tau \to 0$ is as above except that $\digamma$ stays stuck to 1 and the physical quantities are controlled by the Gaussian fixed point. As far as we know, this is the only instance (FBC, $T = T_c$) in which ratios of the Gaussian exponents are measured in FSS and not ratios of MFT exponents. In a similar manner, one may expect that the logarithmic corrections in the Q sector which are due to the scaling field u becoming marginal at d_{uc}, do not show up in the G sector.

Acknowledgments

We would like to thank Yu. Holovatch for his constant support, and for accepting a delay of about a year in the delivery of the present chapter!

This may not be a common practice, but here I would like also to thank two of my collaborators, two PhD students who are working with me, and were under the co-supervision of Ralph. Andy Manapany and Leïla Moueddene. They are wonderful people with whom we can share opinions and views on the world. I think that maybe I opened their minds to some aspects, and I am sure that they opened mine. They knew Ralph, they appreciated him a lot, and we share this.

References

1. B. Berche, T. Ellis, Y. Holovatch, and R. Kenna, Phase transitions above the upper critical dimension, *SciPost Phys. Lect. Notes.* p. 60 (2022). doi: 10.21468/SciPostPhysLectNotes.60. URL `https://scipost.org/10.21468/SciPostPhysLectNotes.60`.

2. M. Fisher, *Notes, definitions, and formulas for critical point singularities.* In Critical phenomena. Proceedings of a conference held in Washington, DC, April 1965, ed. M.S. Green and J.V. Sengers, National Bureau of Standards (1966).

3. R. Kenna, *Universal scaling relations for logarithmic-correction exponents*, In *Order, Disorder and Criticality: Advanced Problems of Phase Transition Theory, Volume 3*, chapter 1, pp. 1–46. World Scientific, Singapore (2012). doi: https://doi.org/10.1142/9789814417891_0001. URL `https://doi.org/10.1142/9789814417891_0001`.

4. R. Kenna, D. Johnston, and W. Janke, Scaling relations for logarithmic corrections, *Phys. Rev. Lett.* **96**, 115701 (March, 2006). doi: 10.1103/PhysRevLett.96.115701. URL `https://doi.org/10.1103/PhysRevLett.96.115701`.

5. R. Kenna, D. Johnston, and W. Janke, Self-consistent scaling theory for logarithmic-correction exponents, *Phys. Rev. Lett.* **97**, 155702 (October, 2006). doi: 10.1103/PhysRevLett.97.155702. URL `https://doi.org/10.1103/PhysRevLett.97.155702`.

6. B. Berche, R. Kenna, and J.-C. Walter, Hyperscaling above the upper critical dimension, *Nuclear Physics B.* **865**(1), 115–132 (2012). ISSN 0550-3213. doi: https://doi.org/10.1016/j.nuclphysb.2012.07.021. URL `https://www.sciencedirect.com/science/article/pii/S0550321312004063`.

7. E. Brézin, An investigation of finite size scaling, *J. Phys. France.* **43**, 15–22 (1982). doi: 10.1051/jphys:0198200430101500.

8. R. Kenna and B. Berche, A new critical exponent 'coppa' and its logarithmic counterpart 'hat coppa', *Condensed Matter Physics.* **16**, 23601:1–12 (2013). doi: 10.5488/CMP.16.23601.

9. R. Kenna and C. B. Lang, Finite-size scaling and the zeroes of the partition function in the ϕ_4^4 model, *Phys. Lett. B.* **264**, 396–400 (1991). doi: https://doi.org/10.1016/0370-2693(91)90367-Y. URL `https://doi.org/10.1016/0370-2693(91)90367-Y`.

10. R. Kenna and C. B. Lang, Renormalization group analysis of finite size scaling in the ϕ_4^4 model, *Nucl. Phys. B.* **393**, 461–479 (1993). doi: https://doi.org/10.1016/0550-3213(93)90068-Z. URL https://doi.org/10.1016/0550-3213(93)90068-Z.

11. R. Kenna and C. B. Lang, Lee-Yang zeroes and logarithmic corrections in the ϕ_4^4 theory, *Nucl. Phys. B (Proc. Suppl.).* **30**, 697–700 (1993). doi: 10.1016/0920-5632(93)90305-P. URL 10.1016/0920-5632(93)90305-P.

12. R. Kenna and C. B. Lang, Scaling and density of Lee-Yang zeroes in the four-dimensional Ising model, *Phys. Rev. E.* **49**, 5012–5017 (1994). doi: https://journals.aps.org/pre/abstract/10.1103/PhysRevE.49.5012. URL https://journals.aps.org/pre/abstract/10.1103/PhysRevE.49.5012.

13. R. Kenna, Finite size scaling for $O(N)$ ϕ^4-theory at the upper critical dimension, *Nucl. Phys. B.* **691**, 292–304 (2004). doi: 10.1016/0920-5632(93)90305-P. URL 10.1016/0920-5632(93)90305-P.

14. B. Widom, Surface tension and molecular correlations near the critical point, *J. Chem. Phys.* **43**, 3892–3897 (1965). doi: 10.1063/1.1696617. URL https://doi.org/10.1063/1.1696617.

15. B. Widom, Equation of state in the neighborhood of the critical point, *J. Chem. Phys.* **43**, 3898–3905 (1965). doi: 10.1063/1.1696618. URL https://doi.org/10.1063/1.1696618.

16. R. Griffiths, Thermodynamic functions for fluids and ferromagnets near the critical point, *Phys. Rev.* **158**, 176–187 (1967). doi: 10.1103/PhysRev.158.176. URL https://doi.org/10.1103/PhysRev.158.176.

17. G. Rushbrooke, Degree of the critical isotherm, *J. Chem. Phys.* **41**, 1633–1635 (1964). doi: https://doi.org/10.1063/1.1726135. URL https://doi.org/10.1063/1.1726135.

18. R. Kenna and B. Berche, Universal finite-size scaling for percolation theory in high dimensions, *Journal of Physics A: Mathematical and Theoretical.* **50**(23), 235001 (may, 2017). doi: 10.1088/1751-8121/aa6bd5. URL https://doi.org/10.1088/1751-8121/aa6bd5.

19. J. Ruiz-Lorenzo, Revisiting (logarithmic) scaling relations using renormalization group, *Condensed Matter Physics.* **20**, 13601 (2017). doi: https://doi.org/10.5488/CMP.20.13601. URL https://doi.org/10.5488/CMP.20.13601.

20. K. Binder, Critical properties and finite-size effects of the five-dimensional Ising model, *Z. Physik B - Condensed Matter.* **61**, 13–23 (1985). doi: 10.1007/BF01308937.

21. R. Kenna and B. Berche, *Scaling and Finite-Size Scaling above the Upper Critical Dimension,* In *Order, Disorder and Criticality: Advanced Problems of Phase Transition Theory, Volume 4,* chapter 1, pp. 1–54. World Scientific, Singapore (2015). doi: 10.1142/9789814632683_0001. URL https://www.worldscientific.com/doi/abs/10.1142/9789814632683_0001.

22. A. Langheld, J. A. Koziol, P. Adelhardt, S. C. Kapfer, and K. P. Schmidt, Scaling at quantum phase transitions above the upper critical dimension, *SciPost Phys.* **13**, 088 (2022). doi: 10.21468/SciPostPhys.13.4.088. URL https://scipost.org/10.21468/SciPostPhys.13.4.088.

23. P. Adelhardt, J. A. Koziol, A. Langheld, and K. P. Schmidt, Monte Carlo based techniques for quantum magnets with long-range interactions, *Entropy.* **26**(5) (2024). ISSN 1099-4300. doi: 10.3390/e26050401. URL `https://www.mdpi.com/1099-4300/26/5/401`.

24. N. G. Fytas, V. Martín-Mayor, G. Parisi, M. Picco, and N. Sourlas, Finite-size scaling of the random-field Ising model above the upper critical dimension, *Phys. Rev. E.* **108**, 044146 (Oct, 2023). doi: 10.1103/PhysRevE.108.044146. URL `https://link.aps.org/doi/10.1103/PhysRevE.108.044146`.

25. I. Bonamassa, B. Gross, J. Kertész, and S. Havlin. Chimeric classes of universality of catastrophic cascades (2024).

26. E. J. Flores-Sola. *Finite-size scaling above the upper critical dimension.* PhD thesis (2016). URL `http://www.theses.fr/2016LORR0337`. Thèse de doctorat dirigée par Berche, Bertrand et Kenna, Ralph Physique Université de Lorraine, Coventry University, 2016.

27. E. Flores-Sola, B. Berche, R. Kenna, and M. Weigel, Role of Fourier modes in finite-size scaling above the upper critical dimension, *Phys. Rev. Lett.* **116**, 115701 (Mar, 2016). doi: 10.1103/PhysRevLett.116.115701. URL `https://link.aps.org/doi/10.1103/PhysRevLett.116.115701`.

28. R. Kenna and B. Berche, Fisher's scaling relation above the upper critical dimension, *EPL.* **105**, 26005 (2014). doi: 10.1209/0295-5075/105/26005. URL `https://iopscience.iop.org/article/10.1209/0295-5075/105/26005`.

29. A. Z. Patashinskii and V. L. Pokrovsky, Behavior of ordered systems near the transition point, *Sov. Phys. JETP.* **23**, 292–297 (1966). URL `http://www.jetp.ac.ru/cgi-bin/e/index/e/23/2/p292?a=list`.

30. L. Kadanoff, Scaling laws for Ising models near T_c, *Physics.* **2**, 263–272 (1966). doi: 10.1103/PhysicsPhysiqueFizika.2.263. URL `https://doi.org/10.1103/PhysicsPhysiqueFizika.2.263`.

31. A. A. V. Privman, P.C. Hohenberg, *Universal Critical-Point Amplitude Relations.* vol. 14, *Phase Transitions and Critical Phenomena, C. Domb and J.L. Lebowitz (Eds.)*, Academic (1991).

32. L. Moueddene, A. Donoso, and B. Berche, Ralph Kenna's scaling relations in critical phenomena, *Entropy.* **26**(3) (2024). ISSN 1099-4300. doi: 10.3390/e26030221. URL `https://www.mdpi.com/1099-4300/26/3/221`.

33. B. Berche, P. Butera, and L. N. Shchur, The two-dimensional 4-state potts model in a magnetic field, *Journal of Physics A: Mathematical and Theoretical.* **46**(9), 095001 (Feb, 2013). doi: 10.1088/1751-8113/46/9/095001. URL `https://dx.doi.org/10.1088/1751-8113/46/9/095001`.

34. J. Essam and M. Fisher, Padé approximant studies of the lattice gas and Ising ferromagnet below the critical point, *J. Chem. Phys.* **38**, 802–812 (1963). doi: 10.1063/1.1733766. URL `https://doi.org/10.1063/1.1733766`.

35. M. Fisher, Correlation functions and the critical region of simple fluids, *J. Math. Phys.* **5**, 944–962 (1964). doi: 10.1063/1.1704197. URL `https://doi.org/10.1063/1.1704197`.

36. B. Josephson, Inequality for the specific heat: I. derivation, *Proc. Phys. Soc.* **92**, 269–275 (1967). doi: 10.1088/0370-1328/92/2/301. URL `https://doi.org/10.1088/0370-1328/92/2/301`.

37. G. Rushbrooke, On the thermodynamics of the critical region for the Ising problem, *J. Chem. Phys.* **39**, 842–843 (1963). doi: 10.1063/1.1734338. URL https://doi.org/10.1063/1.1734338.

38. R. Griffiths, Thermodynamic inequality near the critical point for ferromagnets and fluids, *Phys. Rev. Lett.* **14**, 623–624 (623-624). doi: 10.1103/PhysRevLett.14.623. URL https://doi.org/10.1103/PhysRevLett.14.623.

39. M. Buckingham and J. Gunton, Correlations at the critical point of the Ising model, *Phys. Rev.* **178**, 848–853 (1969). doi: 10.1103/PhysRev.178.848. URL https://doi.org/10.1103/PhysRev.178.848.

40. M. Fisher, Rigorous inequalities for critical-point correlation exponents, *Phys. Rev.* **180**, 594–600 (1969). doi: 10.1103/PhysRev.180.594.

41. D. Nelson. Remembrances of Michael E. Fisher (2023).

42. N. Aktekin, The finite-size scaling functions of the four-dimensional Ising model, *Journal of Statistical Physics.* **104**, 1397–1406 (2001). doi: DOI:10.1023/A:1010457905088. URL DOI:10.1023/A:1010457905088.

43. E. Luijten and H. W. J. Blöte, Classical critical behavior of spin models with long-range interactions, *Phys. Rev. B.* **56**, 8945–8958 (Oct, 1997). doi: 10.1103/PhysRevB.56.8945. URL https://link.aps.org/doi/10.1103/PhysRevB.56.8945.

44. E. Brézin, J. Le Guillou, and J. Zinn-Justin, *Field theoretical approach to critical phenomena*, In eds. C. Domb and M. Green, *Phase transitions and critical phenomena, vol. VI*, pp. 125–247. Academic Press, New York (1976).

45. J. Rudnick, H. Guo, and D. Jasnow, Finite-size scaling and the renormalization group, *J. Stat. Phys.* **41**, 353–373 (Nov, 1985). doi: /10.1007/BF01009013. URL https://doi.org/10.1007/BF01009013.

46. J.-P. Lv, W. Xu, Y. Sun, K. Chen, and Y. Deng, Finite-size scaling of $O(n)$ systems at the upper critical dimensionality, *National Science Review.* **8**(3) (08, 2020). ISSN 2095-5138. doi: 10.1093/nsr/nwaa212. URL https://doi.org/10.1093/nsr/nwaa212. nwaa212.

47. J. L. Jones and A. P. Young, Finite-size scaling of the correlation length above the upper critical dimension in the five-dimensional Ising model, *Phys. Rev. B.* **71**, 174438 (May, 2005). doi: 10.1103/PhysRevB.71.174438. URL https://link.aps.org/doi/10.1103/PhysRevB.71.174438.

48. R. Kenna and B. Berche. On a previously unpublished work with Ralph Kenna (2023).

49. F. Wu, Professor C.N. Yang and statistical mechanics, *International Journal of Modern Physics B.* **22**, 1899–1909 (2008). doi: 10.1142/S0217979208039198. URL https://doi.org/10.1142/S0217979208039198.

50. C. N. Yang and T. D. Lee, Statistical theory of equations of state and phase transitions. I. Theory of condensation, *Phys. Rev.* **87**, 404–409 (Aug, 1952). doi: 10.1103/PhysRev.87.404. URL https://link.aps.org/doi/10.1103/PhysRev.87.404.

51. T. D. Lee and C. N. Yang, Statistical theory of equations of state and phase transitions. II. Lattice gas and Ising model, *Phys. Rev.* **87**, 410–419 (Aug, 1952). doi: 10.1103/PhysRev.87.410. URL https://link.aps.org/doi/10.

1103/PhysRev.87.410.

52. M. E. Fisher, Yang-Lee edge singularity and ϕ^3 field theory, *Phys. Rev. Lett.* **40**, 1610–1613 (Jun, 1978). doi: 10.1103/PhysRevLett.40.1610. URL `https://link.aps.org/doi/10.1103/PhysRevLett.40.1610`.

53. M. Fisher, *The nature of critical points*, In ed. W. E. Brittin, *Lecture Notes in Theoretical Physics*, vol. 7c, pp. 1–159. University of Colorado Press (1965).

54. M. Lüscher and P. Weisz, Scaling laws and triviality bounds in the lattice ϕ^4 theory: (I). one-component model in the symmetric phase, *Nuclear Physics B.* **290**, 25–60 (1987). doi: https://doi.org/10.1016/0550-3213(87)90177-5. URL `https://doi.org/10.1016/0550-3213(87)90177-5`.

55. M. Lüscher and P. Weisz, Scaling laws and triviality bounds in the lattice ϕ^4 theory: (II). one-component model in the phase with spontaneous symmetry breaking, *Nuclear Physics B.* **295**, 65–92 (1989). doi: https://doi.org/10.1016/0550-3213(88)90228-3. URL `https://doi.org/10.1016/0550-3213(88)90228-3`.

56. M. Lüscher and P. Weisz, Scaling laws and triviality bounds in the lattice ϕ^4 theory (III). n-component model, *Nuclear Physics B.* **318**, 705–741 (1989). doi: https://doi.org/10.1016/0550-3213(89)90637-8. URL `https://doi.org/10.1016/0550-3213(89)90637-8`.

57. A. Larkin and D. Khmel'nitskii, Phase transitions in uniaxial ferroelectrics, *Zh. Eksp. Teor. Fiz.* **56**, 2087–2098 (June, 1969). doi: http://jetp.ras.ru/cgi-bin/dn/e_029_06_1123.pdf. URL `http://jetp.ras.ru/cgi-bin/dn/e_029_06_1123.pdf`.

58. A. Deger, K. Brandner, and C. Flindt, Lee-Yang zeros and large-deviation statistics of a molecular zipper, *Phys. Rev. E.* **97**, 012115 (Jan, 2018). doi: 10.1103/PhysRevE.97.012115. URL `https://link.aps.org/doi/10.1103/PhysRevE.97.012115`.

59. A. Deger and C. Flindt, Determination of universal critical exponents using Lee-Yang theory, *Phys. Rev. Research.* **1**, 023004 (Sep, 2019). doi: 10.1103/PhysRevResearch.1.023004. URL `https://link.aps.org/doi/10.1103/PhysRevResearch.1.023004`.

60. A. Deger and C. Flindt, Lee-Yang theory of the Curie-Weiss model and its rare fluctuations, *Phys. Rev. Research.* **2**, 033009 (Jul, 2020). doi: 10.1103/PhysRevResearch.2.033009. URL `https://link.aps.org/doi/10.1103/PhysRevResearch.2.033009`.

61. A. Deger, F. Brange, and C. Flindt, Lee-Yang theory, high cumulants, and large-deviation statistics of the magnetization in the Ising model, *Phys. Rev. B.* **102**, 174418 (Nov, 2020). doi: 10.1103/PhysRevB.102.174418. URL `https://link.aps.org/doi/10.1103/PhysRevB.102.174418`.

62. L. Moueddene, N. G. Fytas, Y. Holovatch, R. Kenna, and B. Berche, Critical and tricritical singularities from small-scale Monte Carlo simulations: the Blume–Capel model in two dimensions, *Journal of Statistical Mechanics: Theory and Experiment.* **2024**(2), 023206 (Feb, 2024). doi: 10.1088/1742-5468/ad1d60. URL `https://dx.doi.org/10.1088/1742-5468/ad1d60`.

63. X. Peng, H. Zhou, B.-B. Wei, J. Cui, J. Du, and R.-B. Liu, Experimental observation of Lee-Yang zeros, *Phys. Rev. Lett.* **114**, 010601 (Jan, 2015).

doi: 10.1103/PhysRevLett.114.010601. URL https://link.aps.org/doi/10.1103/PhysRevLett.114.010601.

64. B.-B. Wei and R.-B. Liu, Lee-Yang zeros and critical times in decoherence of a probe spin coupled to a bath, *Phys. Rev. Lett.* **109**, 185701 (Oct, 2012). doi: 10.1103/PhysRevLett.109.185701. URL https://link.aps.org/doi/10.1103/PhysRevLett.109.185701.

65. N. Ananikian and R. Kenna, Imaginary magnetic fields in the real world, *Physics.* **8** (January, 2015). doi: https://physics.aps.org/articles/v8/2. URL https://physics.aps.org/articles/v8/2.

66. A. Gordillo-Guerrero, R. Kenna, and J. Ruiz-Lorenzo, Universal amplitude ratios in the Ising model in three dimensions, *JSTAT.* p. P09019 (Sep, 2011). doi: https://doi.org/10.1088/1742-5468/2011/09/P09019. URL https://doi.org/10.1088/1742-5468/2011/09/P09019.

67. W. Janke, D. Johnston, and R. Kenna, Phase transition strength through densities of general distributions of zeroes, *Nuclear Physics B.* **682**(3), 618–634 (2004). ISSN 0550-3213. doi: https://doi.org/10.1016/j.nuclphysb.2004.01.028. URL https://www.sciencedirect.com/science/article/pii/S0550321304000562.

68. P. Kopietz, L. Bartosch, and F. Schütz, *Mean-Field Theory and the Gaussian Approximation*, In *Introduction to the Functional Renormalization Group*, pp. 23–52. Springer Berlin Heidelberg, Berlin, Heidelberg (2010). ISBN 978-3-642-05094-7. doi: 10.1007/978-3-642-05094-7_2. URL https://doi.org/10.1007/978-3-642-05094-7_2.

69. M. Blume, Theory of the first-order magnetic phase change in UO_2, *Phys. Rev.* **141**(2), 517–524 (Jan., 1966). doi: 10.1103/PhysRev.141.517. URL https://link.aps.org/doi/10.1103/PhysRev.141.517.

70. H. W. Capel, On the possibility of first-order phase transitions in Ising systems of triplet ions with zero-field splitting, *Physica.* **32**(5), 966–988 (1966). ISSN 0031-8914. doi: https://doi.org/10.1016/0031-8914(66)90027-9. URL https://www.sciencedirect.com/science/article/pii/0031891466900279.

71. M. Blume, V. J. Emery, and R. B. Griffiths, Ising model for the λ transition and phase separation in He^3-He^4 mixtures, *Phys. Rev. A.* **4**(3), 1071–1077 (Sept., 1971). doi: 10.1103/PhysRevA.4.1071. URL https://link.aps.org/doi/10.1103/PhysRevA.4.1071.

72. P. M. Chaikin and T. C. Lubensky, *Principles of Condensed Matter Physics.* Cambridge University Press, Cambridge (1995).

73. M. Fisher, *Scaling, Universality and the Renormalization Group theory*, In *Critical Phenomena*, chapter 1, pp. 1–137. Springer-Verlag Berlin Heidelberg (1983). doi: 110.1007/3-540-12675-9. URL https://www.springer.com/gp/book/9783540126751. Proceedings of the Summer School Held at the University of Stellenbosch, South Africa January 18–29, 1982.

74. M. E. Fisher. Renormalization group in critical phenomena and quantum field theory : Proceedings of a conference, Temple University, May 29-31, 1973. (1974).

75. N. Goldenfeld, *Lectures On Phase Transitions And The Renormalization*

Group. CRC Press (1992). URL `https://doi.org/10.1201/9780429493492`.

76. E. Brézin and J. Zinn-Justin, Finite size effects in phase transitions, *Nuclear Physics B.* **257**, 867–893 (1985). ISSN 0550-3213. doi: https:// doi.org/10.1016/0550-3213(85)90379-7. URL `https://www.sciencedirect.com/science/article/pii/0550321385903797`.

77. C. Rickwardt, P. Nielaba, and K. Binder, A finite size scaling study of the five-dimensional Ising model, *Annalen der Physik.* **506**(6), 483–493 (1994). doi: https://doi.org/10.1002/andp.19945060606. URL `https://onlinelibrary.wiley.com/doi/abs/10.1002/andp.19945060606`.

78. K. Binder, M. Nauenberg, V. Privman, and A. P. Young, Finite-size tests of hyperscaling, *Phys. Rev. B.* **31**, 1498–1502 (Feb, 1985). doi: 10.1103/PhysRevB.31.1498. URL `https://link.aps.org/doi/10.1103/PhysRevB.31.1498`.

79. E. Luijten and H. W. J. Blöte, Finite-size scaling and universality above the upper critical dimensionality, *Phys. Rev. Lett.* **76**, 1557–1561 (Mar, 1996). doi: 10.1103/PhysRevLett.76.1557. URL `https://link.aps.org/doi/10.1103/PhysRevLett.76.1557`.

80. E. Luijten and H. W. J. Blöte, Finite-size scaling and universality above the upper critical dimensionality, *Phys. Rev. Lett.* **76**, 3662–3662 (May, 1996). doi: 10.1103/PhysRevLett.76.3662.3. URL `https://link.aps.org/doi/10.1103/PhysRevLett.76.3662.3`.

81. M. Wittmann and A. P. Young, Finite-size scaling above the upper critical dimension, *Phys. Rev. E.* **90**, 062137 (Dec, 2014). doi: 10.1103/PhysRevE.90.062137. URL `https://link.aps.org/doi/10.1103/PhysRevE.90.062137`.

82. E. Luijten. *Interaction Range, Universality and the Upper Critical Dimension.* PhD thesis (1997). URL `http://resolver.tudelft.nl/uuid:8ec8fc17-4a7e-47e3-aec0-bceedcfa5904`. Doctoral thesis, Delft University Press.

83. J. Grimm, E. M. Elçi, Z. Zhou, T. M. Garoni, and Y. Deng, Geometric explanation of anomalous finite-size scaling in high dimensions, *Phys. Rev. Lett.* **118**, 115701 (Mar, 2017). doi: 10.1103/PhysRevLett.118.115701. URL `https://link.aps.org/doi/10.1103/PhysRevLett.118.115701`.

84. J. F. Nagle and J. C. Bonner, Numerical studies of the Ising chain with long-range ferromagnetic interactions, *Journal of Physics C: Solid State Physics.* **3**(2), 352–366 (Feb, 1970). doi: 10.1088/0022-3719/3/2/017. URL `https://doi.org/10.1088/0022-3719/3/2/017`.

85. G. A. Baker and G. R. Golner, Spin-spin correlations in an Ising model for which scaling is exact, *Phys. Rev. Lett.* **31**, 22–25 (Jul, 1973). doi: 10.1103/PhysRevLett.31.22. URL `https://link.aps.org/doi/10.1103/PhysRevLett.31.22`.

86. V. Papathanakos, Finite-size effects in high-dimensional statistical mechanical systems: The Ising model with periodic boundary conditions, *PhD thesis, Princeton University* (November, 2006).

87. K. Binder and E. Luijten, Monte Carlo tests of theoretical predictions for critical phenomena: still a problem?, *Computer Physics Communications.* **127**(1), 126–130 (2000). ISSN 0010-4655. doi: https://doi.org/10.1016/

S0010-4655(00)00015-1. URL `https://www.sciencedirect.com/science/article/pii/S0010465500000151`.

88. B. D. Josephson, Inequality for the specific heat: I. Derivation, *Proceedings of the Physical Society.* **92**(2), 269 (Oct, 1967). doi: 10.1088/0370-1328/92/2/301. URL `https://dx.doi.org/10.1088/0370-1328/92/2/301`.

89. Z. Merdan and D. Gokbel-Keklikoglu, The test of a new critical exponent by using Ising model on the creutz cellular automaton, *Acta Physica Polonica A.* **133**, 1200–1204 (January, 2018). doi: 10.12693/APhysPolA.133.1200. URL `http://przyrbwn.icm.edu.pl/APP/PDF/133/app133z5p15.pdf`.

90. P. Lundow and K. Markström, Finite size scaling of the 5d Ising model with free boundary conditions, *Nuclear Physics B.* **889**, 249–260 (2014). ISSN 0550-3213. doi: https://doi.org/10.1016/j.nuclphysb.2014.10.011. URL `https://www.sciencedirect.com/science/article/pii/S0550321314003095`.

91. M. C. Wittmann. *Nature of the spin-glass phase in models with long-range interactions.* PhD thesis (2015). URL `https://escholarship.org/uc/item/4cx4n67h#author`. Doctoral thesis, UC Santa Cruz.

92. J. Rudnick, G. Gaspari, and V. Privman, Effect of boundary conditions on the critical behavior of a finite high-dimensional Ising model, *Phys. Rev. B.* **32**, 7594–7596 (Dec, 1985). doi: 10.1103/PhysRevB.32.7594. URL `https://link.aps.org/doi/10.1103/PhysRevB.32.7594`.

93. Y. Honchar, B. Berche, Y. Holovatch, and R. Kenna, When correlations exceed system size: finite-size scaling in free boundary conditions above the upper critical dimension, *Condensed Matter Physics.* **27**(1), 13603 (2024). doi: 10.5488/CMP.27.13603.

Chapter 2

Stochastic Spatial Lotka–Volterra Predator-Prey Models

Uwe C. Täuber*

*Department of Physics & Center for Soft Matter and Biological Physics,
850 West Campus Drive (MC 0435), Faculty of Health Sciences,
Virginia Tech, Blacksburg, VA 24061, USA*

Dynamical models of interacting populations are of fundamental interest for spontaneous pattern formation and other noise-induced phenomena in nonequilibrium statistical physics. Theoretical physics in turn provides a quantitative toolbox for paradigmatic models employed in (bio-)chemistry, biology, ecology, epidemiology, and even sociology. Stochastic, spatially extended models for predator-prey interaction display spatio-temporal structures that are not captured by the Lotka–Volterra mean-field rate equations. These spreading activity fronts reflect persistent correlations between predators and prey that can be analyzed through field-theoretic methods. Introducing local restrictions on the prey population induces a predator extinction threshold, with the critical dynamics at this continuous active-to-absorbing state transition governed by the scaling exponents of directed percolation. Novel features in biologically motivated model variants include the stabilizing effect of a periodically varying carrying capacity that describes seasonally oscillating resource availability; enhanced mean species densities and local fluctuations caused by spatially varying reaction rates; and intriguing evolutionary dynamics emerging when variable interaction rates are affixed to individuals combined with trait inheritance to their offspring. The basic susceptible-infected-susceptible and susceptible-infected-recovered models for infectious disease spreading near their epidemic thresholds are respectively captured by the directed and dynamic isotropic percolation universality classes. Systems with three cyclically competing species akin to spatial rock-paper-scissors games may display striking spiral patterns, yet conservation laws can prevent such noise-induced structure formation. In diffusively coupled inhomogeneous settings, one may observe the stabilization of vulnerable ecologies prone to finite-size extinction or fixation due to immigration waves emanating from the interfaces.

---

*tauber@vt.edu, ORCID: https://orcid.org/0000-0001-7854-2254

Contents

1. Introduction: Stochastic spatial population dynamics 68
2. Fluctuations and correlations in reacting particle systems 70
    2.1. Stochastic kinetics and master equation 71
    2.2. Mean-field rate equations 72
    2.3. Individual-based Monte Carlo simulations 74
3. Predator-prey competition and coexistence 75
    3.1. Lotka–Volterra model and population oscillations 76
    3.2. Stochastic lattice model: Noise-induced activity fronts 78
    3.3. Renormalized reaction rates and oscillation parameters 81
    3.4. Predator extinction: Critical dynamics and aging 84
4. Biologically motivated model variants and extensions 88
    4.1. Periodically varying carrying capacity 89
    4.2. Fitness enhancement through environmental variability 91
    4.3. Demographic variability, hereditary trait optimization 93
5. Stochastic spatial models for infectious disease spreading 96
    5.1. SIS model and directed percolation 96
    5.2. SIR model and dynamic isotropic percolation 97
6. Cyclic dominance of three competing species 99
    6.1. Rock-paper-scissors or cyclic Lotka–Volterra model 100
    6.2. Spiral structure formation in the May–Leonard model 102
    6.3. Inhomogeneous systems: diffusively coupled patches 103
7. Conclusions and outlook 107
References 109

1. Introduction: Stochastic spatial population dynamics

The study of living organisms, which per definition constitute complex dynamical systems comprising many interacting individual constituents, has become a vast and fertile research arena for physicists and mathematicians trained in nonequilibrium statistical mechanics and the theory of stochastic processes. Comparing with the more traditional applications of statistical physics to condensed matter systems, one may observe many common but also distinct features in biological applications. While the number of degrees of freedom for biological systems is certainly sufficiently large to render a statistical description meaningful and productive, it is also typically much smaller than Avogadro's number. Hence statistical fluctuations are usually non-negligible, become amplified owing to their being situated far away from thermal equilibrium, and are often driven by random external influences. Fluctuations are increasingly acknowledged to importantly contribute to biological functionality. Consequently the thermodynamic limit may not apply, merely studying average quantities is insufficient, finite-size effects are crucial, and transient phenomena dominate any living entity.

Biological structures as well as dynamical processes are also characteristically optimized due to evolutionary mechanisms. Thus both spatial and temporal correlations play prominent roles, manifesting themselves in spontaneous structure formation and the presence of history dependence and memory effects. In addition, as opposed to atoms or molecules, cells and individuals in a population are not identical, but carry specific traits that affect their dynamics and may also change over time. Nevertheless, as a theoretical physicist one hopes – almost in defiance of the astounding complexity of the biological realm, and despite the abundant difficulties in separating clearly distinct length and time scales that is so profoundly important in establishing well-defined effective description levels – that exhaustive studies of comparatively simple fundamental models that employ the full, powerful analytical and computational machinery developed in statistical physics may contribute a deeper understanding of both structural and dynamical features encountered in the living world.

This chapter addresses exemplary basic models in ecology and epidemiology that share a common mathematical language with chemical reactions, namely stochastic spatially extended systems that are dominated by a binary "predation" or "infection" process $A + B \to A + A$, where upon mutual encounter an individual or particle of "prey" or "susceptible" species B becomes replaced with a "predator" or "infectious" individual A, with a prescribed rate λ that will serve as the pertinent nonlinear control parameter.[1–3] The corresponding mean-field rate equations are the celebrated coupled ordinary differential equations independently proposed about a century ago by Alfred Lotka and Vito Volterra who aimed to mathematically capture oscillatory dynamics in the quite distinct contexts of autocatalytic chemical reactions[4] and fish populations in the Adriatic sea.[5]

Indeed, much of the traditional original and textbook literature on chemical kinetics,[1,2] population ecology,[2,3,6,7] and closely related evolutionary game theory[8,9] invokes deterministic chemical rate equations or their spatial extension, reaction-diffusion models, which constitute mean-field type approximations that neglect stochastic fluctuations and spatio-temporal correlations through "mass action" factorizations of higher moments, see Sec. 2.[10–13] With the dramatic enhancement of computational power since the 1970s and the accompanying development of versatile analytical tools, stochastic models for (bio-)chemical reaction kinetics, population dynamics, and infectious spreading have become an increasingly active research field in nonequilibrium statistical physics, with the focus often on relevant fluctuation and correlation effects that are not captured by their representation in terms of mere rate equations, *e.g.* Refs. [10–26] (this is merely a non-exhaustive list of reviews, recent monographs, and textbooks; please also see the extensive references cited therein).

As described in Sec. 3, Lotka–Volterra predator-prey systems display many non-trivial phenomena that are strongly influenced by intrinsic dynamical correlations and fluctuations: The species coexistence regime is dominated by persistent activity waves traversing the system where surviving predators necessarily pursue prey that expand into unoccupied space. These noise-generated and -stabilized correlated spatio-temporal structures in turn induce stochastic and resonantly amplified population oscillations whose quantitative features are markedly renormalized relative to mean-field theory. Through a nonlinear feedback mechanism, limited prey resources encoded in a finite local carrying capacity severely affect the predator population, even causing its extinction and thus prey fixation at a critical threshold $\lambda < \lambda_c$. The ensuing stationary critical dynamics as well as universal transient aging features that characterize the far-from-equilibrium continuous phase transition from the active coexistence to the inactive, absorbing predator extinction state are governed by the scaling exponents of the generic directed percolation universality class.

The biological context moreover suggests model variations that would not usually be considered in a chemical or physical setting; some examples are discussed in Sec. 4, namely periodic changes in the environment and reaction rates that are specific to individual particles and whose distribution may evolve under Darwinian competition with mutations. Subsequently, Secs. 5 and 6 respectively provide succinct overviews of the stochastic spatial dynamics observed in the closely related fundamental susceptible-infected-susceptible and susceptible-infected-recovered models for the spreading of infectious diseases, specifically their near-threshold critical dynamics, as well as spontaneous spiral pattern formation or the absence thereof in the May–Leonard and "rock-paper-scissors" model variants designed to describe cyclic dominance of three competing species. The chapter concludes (Sec. 7) with the author's brief perspective and outlook.

2. Fluctuations and correlations in reacting particle systems

We begin with a concise outline of the mathematical description of stochastic reacting particle systems through master equations, highlighting the mean-field factorization approximation that is invoked to arrive at the associated coupled rate equations. The standard setup of Monte Carlo simulation algorithms to statistically sample observables is introduced as well.

2.1. *Stochastic kinetics and master equation*

"Chemical reactions" denote (stochastic) processes where through mutual interactions particles representing molecules of certain chemical species, nuclei or elementary particles, individuals in a population, *etc.* alter their identity. For example, consider the reversible binary reaction

$$A + B \underset{\sigma}{\overset{\lambda}{\rightleftharpoons}} C \,, \tag{1}$$

where the notation "$A + B$" on the left-hand side implicitly expresses the condition that one particle each of species A and B must meet at the same location and time in order for the "forward" reaction with rate λ to occur; in contrast, the reverse reaction with rate σ may happen spontaneously as long as any C particles remain present. The configuration of the system at any time t is fully characterized by specifying the (local) number of particles $n_\alpha = 0, 1, \ldots$ of each species $\alpha = A, B, C, \ldots$ at that instant. Since reactions just modify these integers $\{n_\alpha\}$ by the associated stoichiometric coefficients, they represent Markovian stochastic processes in species number space. In example (1), the reaction rates λ and σ then determine the associated configuration-dependent transition rates

$$w\big(\{n_A, n_B, n_C\} \to \{n_A - 1, n_B - 1, n_C + 1\}\big) = \lambda\, n_A n_B \,,$$
$$w\big(\{n_A, n_B, n_C\} \to \{n_A + 1, n_B + 1, n_C - 1\}\big) = \sigma\, n_C \,, \tag{2}$$

as there are n_α possibilities to pick one of the reactant particles of type α.

To generalize the above local processes to spatially extended systems, one merely needs to view each species α on distinct lattice or continuum positions i as "quasi-species" that are fully characterized by integer occupation numbers $n_{\alpha i}$ which are in turn altered by integer values through reactions and / or transport. Unrestricted hopping of a single particle from site i to j with rate D is then governed by the Markovian transition rate

$$w\big(\{n_{\alpha i}, n_{\alpha j}\} \to \{n_{\alpha i} - 1, n_{\alpha j} + 1\}\big) = D\, n_{\alpha i} \,. \tag{3}$$

Yet if, for example, at most only a single particle may occupy any location i, so that all $n_{\alpha i} = 0$ or 1, the right-hand side would need to be multiplied with the factor $(1 - n_{\alpha j})$, rendering the resulting exclusion process non-linear in nature. In biology, site restrictions that constrain local particle occupations may be caused by limited resources for the affected species, and the maximum sustained number is referred to as "carrying capacity".[2,3]

The "chemical" master equation constitutes a deterministic evolution equation for the configurational probability $P(\{n_\alpha\};t)$ that accounts for the gain-loss balance due to reactive transitions,[10,27,28]

$$\frac{\partial P(\{n_\alpha\};t)}{\partial t} = \sum_{\{n'_\alpha \neq n_\alpha\}} \Big[P(\{n'_\alpha\};t)\, w(\{n'_\alpha\} \to \{n_\alpha\};t)$$

$$-P(\{n_\alpha\};t)\, w(\{n_\alpha\} \to \{n'_\alpha\};t) \Big] . \qquad (4)$$

In our example (1) governed by the rates (2),

$$\frac{\partial P(n_A, n_B, n_C;t)}{\partial t} = \lambda\,(n_A + 1)\,(n_B + 1)\,P(n_A + 1, n_B + 1, n_C - 1;t)$$

$$+\sigma\,(n_C + 1)\,P(n_A - 1, n_B - 1, n_C + 1;t)$$

$$- (\lambda\,n_A n_B + \sigma n_C)\,P(n_A, n_B, n_C;t) , \qquad (5)$$

with the convention that $P(\{n_\alpha\};t) = 0$ if any $n_\alpha < 0$. In terms of the varying particle numbers, Eqs. (5) represent an infinite set of coupled linear ordinary differential equations. Any observable quantity O needs to be a function of the particle numbers $\{n_\alpha\}$, and hence its expectation value with respect to the ensemble of possible configurations is

$$\langle O(\{n_\alpha\},t) \rangle = \sum_{\{n_\alpha\}} O(\{n_\alpha\})\, P(\{n_\alpha\};t) , \qquad (6)$$

with its temporal evolution instilled by the master equation (4) for the probabilities $P(\{n_\alpha\};t)$, which in principle allows us to compute characteristic time-dependent expectation values, moments, and correlation functions.

2.2. *Mean-field rate equations*

To illustrate this process, consider the average particle numbers,

$$\frac{\partial \langle n_\alpha(t) \rangle}{\partial t} = \sum_{\{n_\alpha\}} n_\alpha \frac{\partial P(\{n_\alpha\};t)}{\partial t} . \qquad (7)$$

As the infinite particle number sums range over all allowed integers, one may simply shift internal summation indices in the master equation gain terms to rewrite them as multiplying $P(\{n_\alpha\};t)$; after resulting cancellations, the remainder can be expressed as other expectation values or moments.[10,11,13] For example, with Eq. (5) one arrives at

$$R(t) = \frac{\partial \langle n_{A/B}(t) \rangle}{\partial t} = -\frac{\partial \langle n_C(t) \rangle}{\partial t} = -\lambda\,\langle [n_A n_B](t) \rangle + \sigma\,\langle n_C(t) \rangle , \qquad (8)$$

where the identities for the net reaction rate $R(t)$ on the left-hand side follow directly from the scheme (1); in general stoichiometric coefficients for the reactions would enter here. In the asymptotic long-time steady state, the average particle numbers become constant, and the exact relationship $\langle n_C(\infty)\rangle/\langle [n_A n_B](\infty)\rangle = \lambda/\sigma$ holds. Yet, in order to compute $\langle n_\alpha(t)\rangle$ for a reaction scheme involving binary or higher-order reactions, one requires the time evolution of nonlinear moments, which iteratively generates an infinite hierarchy of linear coupled differential equations for all moments or correlators. Closure towards a finite set of equations can only be achieved by assuming certain high- order correlations to be insignificant and applying factorizations of the associated moments in products of lower-order ones.[12]

The simplest, but also most drastic approximation is to neglect any two-point and higher- order correlations, which in our above example amounts to the mean-field decoupling $\langle [n_A n_B](t)\rangle \approx \langle n_A(t)\rangle \langle n_B(t)\rangle$ that should be applicable when the reactants are maintained well-mixed in the system, or for reactions in dilute gases or solutions with abundant particle numbers where any spatial and temporal fluctuations are very small compared to the mean reactant densities. Consequently, the reaction rate (8) reduces to $R(t) \approx -\lambda \langle n_A(t)\rangle \langle n_B(t)\rangle + \sigma \langle n_C(t)\rangle$, resulting in a closed set of three coupled (two of which are identical) but now nonlinear ordinary differential equations for the mean particle numbers. These rate equations may subsequently be analyzed by the standard methods of nonlinear dynamics.[1,2] For the stationary particle number averages, the mean-field approximation yields the ratio $\langle n_A(\infty)\rangle \langle n_B(\infty)\rangle/\langle n_C(\infty)\rangle = \sigma/\lambda$ which is just the chemical "law of mass action" applied to the reversible reactions (1) in the dilute reactant limit; in thermal equilibrium at temperature T, the logarithm of this backward-to-forward reaction rate ratio (the "pK value") can furthermore be identified as the net reaction enthalpy relative to $k_B T$.[28,29] In order to incorporate spatial variations, one may supplement the rate equations by, say, diffusive spreading terms $\sim D \nabla^2 n_\alpha(x,t)$ for the coarse-grained local particle densities, which generates "reaction-diffusion" equations that are however still subject to the mass action factorization assumptions.[2]

More refined mean-field theories apply the factorization approximation to higher-order correlations.[18,22] A systematic approach to faithfully include "demographic" fluctuations or "internal reaction noise" and correlations as well as to properly account for the fundamentally discrete nature of the stochastic particle interactions rests on a bosonic Fock state representation of the basic linear master equation kinetics and a subsequent construction of coherent-state path integrals in the continuum limit.[10,19,30–33]

The resulting Doi–Peliti field theory may subsequently be analyzed perturbatively, with the associated "classical field equations" recovering the mean-field reaction-diffusion equations.[34,35] For at most binary reactions, the field theory action can be "reversely" mapped to stochastic Langevin partial differential equations that incorporate multiplicative noise.[10,34–38] Moreover, universal scale-invariant features can be extracted by means of dynamic renormalization group methods[10,19,25] (please also refer to the original citations therein).

2.3. *Individual-based Monte Carlo simulations*

An efficient numerical tool to investigate fluctuation and correlation effects in (bio-)chemical and physical reactions or ecological and epidemiological dynamics is to implement stochastic "individual-" or "agent-based" Monte Carlo simulations.[13] These represent an underlying Markov chain master equation, since any configurational updates only depend on the immediate past, with the corresponding transition rates prescribed in the same manner as in the examples (2) and (3). Stochastic reaction, hopping, or other spreading processes can be implemented on regular lattices, often with periodic boundary to eliminate boundary effects, or appropriately selected contact networks. A typical simulation algorithm proceeds as follows: The system is initialized by distributing particles of either species among available sites, typically randomly, but constrained by specified local occupation number restrictions, if applicable. Next, an individual is selected at random, and depending on the presence of other particles on its site i or in its vicinity, possible reactions are executed with their specified probabilities and subject to any pertinent conditions. For example, pair processes require two particles to meet at the same or neighboring sites; and offspring production to an adjacent location may be restricted by a finite local carrying capacity. Note that in exclusion models with $n_{\alpha i} = 0$ or 1 only, these reactions must necessarily be implemented "off-site", and birth processes hence inevitably generate spatial spreading even in the absence of explicit particle hopping or exchange processes. The particle selection and reaction steps are then repeated $N = \sum_{\alpha,i} n_{\alpha i}$ times such that on average each individual present in the system at this instant has been subject to one of the possible stochastic processes; this completes one Monte Carlo step (MCS) in the sequence of random updates.

Physical time is assumed to be proportional to this artificial simulation clock time. The algorithm may be run sufficiently long to reach a (quasi-)stationary configuration. To control statistical errors, adequate averages must subsequently be taken over many independent initializations and stochastic histories; these may be interpreted as direct sampling of observables subject to the time-dependent configurational probabilities evolving according to the corresponding master equation. Examples are the temporal evolution of mean particle numbers or appropriate correlations that provide quantitative mathematical insights; individual runs may be visualized as illustrative simulation movies. Properly sampled Fourier transforms of the time tracks may be utilized to detect characteristic frequencies and attenuation time scales of associated oscillations through Fourier spectrum peak locations and widths.

Of course, there are inevitable artifacts that manifest themselves on short time and length scales, originating in somewhat arbitrary choices in the algorithmic setups.[13] Yet the goal is to capture generic, perhaps emergent dynamical features on meso- or macroscopic coarse-grained scales that do not crucially depend on microcopic details. One should thus in principle execute distinct algorithms, and also check for the prominence of finite-size effects, *e.g.*, when emerging dynamical structures reach across the entire system. It is also worth noting that in any case, macroscopic system parameters such as effective reaction rates are usually not directly and simply connected with the microscopic transition probabilities implemented in the computer code but may acquire nontrivial renormalizations due to intrinsic dynamical correlations.[39]

3. Predator-prey competition and coexistence

This section discusses the simplest stochastic spatially extended model for predator-prey competition and coexistence which reduces to the famed Lotka–Volterra equations in the mean-field approximation. Predators A and prey B by themselves are subject to linear death $A \xrightarrow{\mu} \emptyset$ and birth (asexual reproduction) processes $B \xrightarrow{\sigma} B + B$. Here, one may also interpret the parameters $-\mu$ and σ as the net reproduction (difference of birth and death) rates for either species, which would respectively result in exponential decay $\langle n_A(t) \rangle = \langle n_A(0) \rangle e^{-\mu t}$ and growth $\langle n_B(t) \rangle = \langle n_B(0) \rangle e^{\sigma t}$. Competition and population control is achieved through the binary predation reaction $A + B \xrightarrow{\lambda} A + A$: Upon mutual encounter, a prey individual becomes replaced with a predator with rate λ.

3.1. *Lotka–Volterra model and population oscillations*

The master equation associated with the above stochastic predator death, prey birth, and predation processes reads

$$
\frac{\partial P(n_A, n_B; t)}{\partial t} = \mu\,(n_A + 1)\,P(n_A + 1, n_B; t) + \sigma\,(n_B - 1)\,P(n_A, n_B - 1; t)
$$
$$
+ \lambda\,(n_A - 1)\,(n_B + 1)\,P(n_A - 1, n_B + 1; t)
$$
$$
- (\mu\,n_A + \sigma\,n_B + \lambda\,n_A n_B)\,P(n_A, n_B; t)\,. \tag{9}
$$

Applying the mass action factorization approximation $\langle [n_A n_B](t) \rangle \approx \langle n_A(t) \rangle \langle n_B(t) \rangle$ then yields the original deterministic Lotka–Volterra model in the form of two coupled ordinary differential equations,

$$
\frac{\partial \langle n_A(t) \rangle}{\partial t} = -\mu\,\langle n_A(t) \rangle + \lambda\,\langle n_A(t) \rangle \langle n_B(t) \rangle\,,
$$
$$
\frac{\partial \langle n_B(t) \rangle}{\partial t} = \sigma\,\langle n_B(t) \rangle - \lambda\,\langle n_A(t) \rangle \langle n_B(t) \rangle\,. \tag{10}
$$

These mean-field rate equations permit three distinct stationary solutions: (i) Total population extinction $\langle n_A(\infty) \rangle = 0 = \langle n_B(\infty) \rangle$; (ii) another absorbing state with predator extinction $\langle n_A(\infty) \rangle = 0$, but Malthusian prey population explosion $\langle n_B(t) \rangle \to \infty$; and (iii) predator-prey coexistence with finite $\langle n_A(\infty) \rangle = \sigma/\lambda$, $\langle n_B(\infty) \rangle = \mu/\lambda$. The predators of course benefit from high prey fertility σ, whereas the prey strive under conditions of high predator mortality μ and low predation efficacy λ; yet perhaps counter-intuitively, increased λ values cause a reduction in the predator population, too, due to a nonlinear feedback mechanism: Frequent predation events deplete the prey, leaving only scarce food recources for the predators.

The predator-prey coexistence state (iii) is however never reached under the mean-field dynamics (10) if the initial configuration is even slightly displaced from it: For small deviations $\delta_{A/B}(t) = \langle n_{A/B}(t) \rangle - \langle n_{A/B}(\infty) \rangle$, one may linearize the equations of motion to obtain $\partial \delta_A(t)/\partial t \approx \sigma\,\delta_B(t)$, $\partial \delta_B(t)/\partial t \approx -\mu\,\delta_A(t)$, which combine to the harmonic oscillator differential equation $\partial^2 \delta_{A/B}(t)/\partial t^2 \approx -\mu\,\sigma\,\delta_{A/B}(t)$ with the oscillation frequency $\omega_0 = \sqrt{\mu\,\sigma}$. Equivalently, one may construct the linear stability matrix

$$
\mathbf{L} = \begin{pmatrix} 0 & \sigma \\ -\mu & 0 \end{pmatrix}
$$

that dictates the time evolution $\partial \mathbf{v}(t)/\partial t \approx \mathbf{L}\,\mathbf{v}(t)$ of the fluctuation vector $\mathbf{v} = (\delta_A\ \delta_B)^T$. The eigenvalues $\pm i\omega_0$ of $\mathbf{L}$ are purely imaginary, indicating undamped oscillatory kinetics in phase space about the fixed point (iii). These neutral cycles are indeed confirmed via eliminating time from

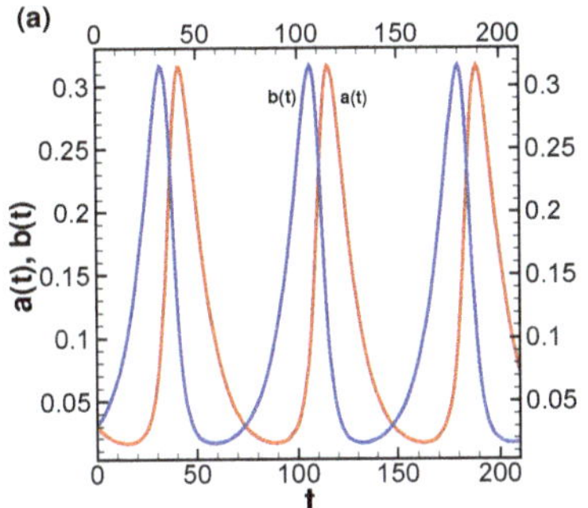
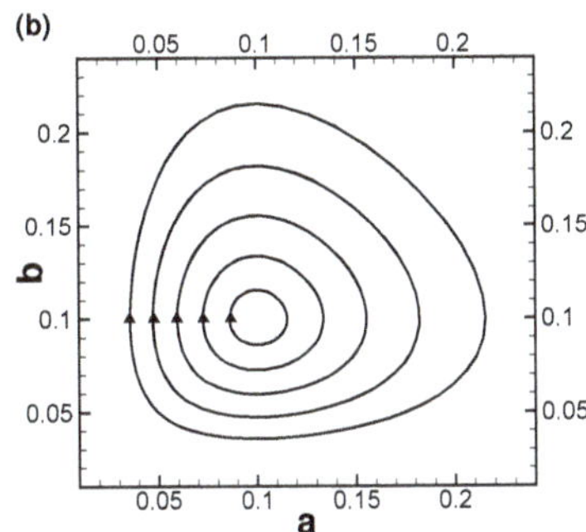

Figure 1. (a) Predator (red) and prey (blue) population oscillations resulting from the deterministic Lotka–Volterra rate equations (10): $\sigma = 0.1$, $\mu = 0.1$, $\lambda = 1$. (b) Periodic orbits in population number phase space. [Figures reproduced with permission from Ref. [53], copyright (2007) by IOP Publ.]

Eqs. (10): $d\langle n_A \rangle / d\langle n_B \rangle = [(\lambda \langle n_B \rangle - \mu) \langle n_A \rangle] / [(\sigma - \lambda \langle n_A \rangle) \langle n_B \rangle]$, which yields, after variable separation and integration, the conserved first integral (Lyapunov function) for the mean-field dynamics $l(t) = \lambda \langle n_A(t) \rangle + \lambda \langle n_B(t) \rangle - \sigma \ln \langle n_A(t) \rangle - \mu \ln \langle n_B(t) \rangle = l(0)$. Consequently, the trajectories in the population number phase space must be periodic orbits, shown in Fig. 1, that inevitably return to the initial configuration $\langle n_A(0) \rangle$, $\langle n_B(0) \rangle$.

However, the conservation of the quantity l only holds within mean-field theory, and is not valid for the original stochastic reaction processes. Moreover, the fine-tuned vanishing real parts in the stability eigenvalues indicate a non-generic approximation artifact that should not remain robust as the model itself is altered. For example, one may implement a finite prey carrying capacity K to prevent their population to diverge in case the predators go extinct. On the rate equation description level, this is captured by a logistic modification of the prey reproduction term,

$$\frac{\partial \langle n_B(t) \rangle}{\partial t} = \sigma \langle n_B(t) \rangle \left[1 - \frac{\langle n_B(t) \rangle}{K} \right] - \lambda \langle n_A(t) \rangle \langle n_B(t) \rangle . \qquad (11)$$

The ensuing modified stationary states are (ii') $\langle n_A(\infty) \rangle = 0$, $\langle n_B(\infty) \rangle = K$ and (iii') $\langle n_A(\infty) \rangle = (\sigma/\lambda) [1 - \mu/(\lambda K)]$, $\langle n_B(\infty) \rangle = \mu/\lambda$, which exists for $\lambda > \lambda_c = \mu/K$ and is then linearly stable. For $\lambda < \lambda_c$, the predator species is driven to extinction (ii'), while the prey population fixates at the carrying capacity K. At the coexistence fixed point (iii'), the eigenvalues of the stability matrix

$$\mathbf{L} = \begin{pmatrix} 0 & \sigma \left(1 - \mu/\lambda K \right) \\ -\mu & -\mu \sigma/\lambda K \end{pmatrix}$$

                                    U. C. Täuber

acquire negative real parts:

$$\epsilon_{\pm} = -\frac{\mu\,\sigma}{2\,\lambda\,K}\left[1 \pm \sqrt{1 - \frac{4\,\lambda\,K}{\sigma}\left(\frac{\lambda\,K}{\mu} - 1\right)}\,\right]. \tag{12}$$

For $\sigma \geq \bar{\sigma} = 4\,\lambda\,K\,(\lambda\,K/\mu - 1)$, both eigenvalues are real, indicating direct exponential relaxation towards the stable node (iii'); on the other hand, if $\sigma < \bar{\sigma}$, the phase space trajectories spiral inwards to reach the stable focus (iii'), with the imaginary part of $\epsilon_{\pm}$ yielding the frequency of the resulting damped population oscillations.

The well-mixed rate equations can be amended to allow for spatial structures by replacing the mean population numbers with local density fields $n_{A/B}(x,t)$ (where $\langle n_{A/B}(t)\rangle = \langle n_{A/B}(x,t)\rangle$) and adding diffusion terms,

$$\frac{\partial n_A(x,t)}{\partial t} = D_A\nabla^2 n_A(x,t) - \mu\,n_A(x,t) + \lambda\,n_A(x,t)\,n_B(x,t)\,, \tag{13}$$

$$\frac{\partial n_B(x,t)}{\partial t} = D_B\nabla^2 n_B(x,t) + \sigma\,n_B(x,t) - \frac{\sigma}{K}n_B(x,t)^2 - \lambda\,n_A(x,t)n_B(x,t)\,.$$

In one dimension, these coupled partial differential equations support traveling wave solutions describing predator fronts that originate in a region set in the coexistence state (iii') and invade prey-occupied space.[2]

3.2. *Stochastic lattice model: Noise-induced activity fronts*

Monte Carlo simulations on sufficiently large two- and three-dimensional regular lattices (with periodic boundary conditions) qualitatively confirm several salient features predicted by the mean-field rate equations, but also display marked differences.[21,26,40–54] As shown in Fig. 2(a), prominent population oscillations are observed in the predator-prey coexistence state. Yet their properties are actually independent of the initial configurations and densities, at variance with the neutral centers resulting from Eqs. (10); indeed the quantity $l(t)$ is not constant in time, and stochastic fluctuations induce small damping in the resulting oscillatory modes. The population oscillation amplitudes are seen to diminish with increasing system size.

Upon implementing site occupation number restrictions, allowing at most a single particle of either species on each lattice point (*i.e.*, $n_{A/B\,i} = 0, 1$), one observes three distinct regimes akin to the analysis of the Lotka–Volterra rate equations with finite prey carrying capacity Eq. (11), depicted in Fig. 2(b): Predator extinction and prey fixation for small predation rates $\lambda < \lambda_c$, albeit with a critical point location that is not in agreement with the mean-field prediction; a node-like two-species coexistence fixed point

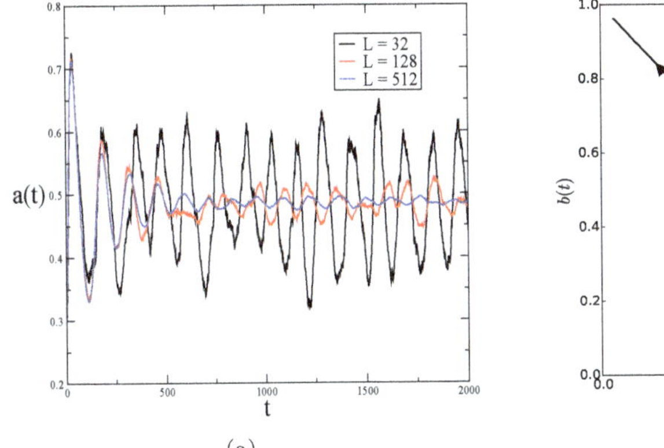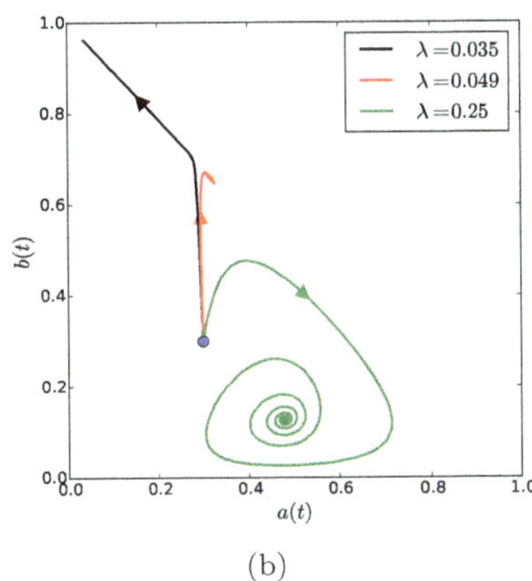

Figure 2.   (a) Predator population oscillations (single Monte Carlo simulation runs) in a stochastic Lotka–Volterra model with site exclusion ($n_{A/B\,i} = 0, 1$) on $L \times L$ square lattices (periodic boundary conditions) of linear system sizes $L = 32$, 128, and 512; initial predator and prey densities 0.3; $\mu = 0.025$, $\sigma = 1$, $\lambda = 0.25$; $D_{A/B} = 0$. The amplitude of the (damped) stochastic oscillations for the mean particle density decreases with L. (b) Typical phase space trajectories for single Monte Carlo runs on a 512×512 square lattice with $\mu = 0.025$, $\sigma = 1$, $D_{A/B} = 0$, and different predation rates, corresponding to the predator extinction / prey fixation state ($\lambda = 0.0035$), and two distinct predator-prey coexistence fixed points, either nodal ($\lambda = 0.049$) or focal ($\lambda = 0.25$). Near the latter, deep in the coexistence regime, one observes resonantly amplified oscillatory population fluctuations. [Figures reproduced with permission from (a) Ref. [21], copyright (2007) by Springer Science+Business Media, Inc.; (b) Ref. [26], copyright (2018) by IOP Publ.]

with exponential relaxation of the mean population numbers towards their steady-state values beyond the threshold $\lambda > \lambda_c$; and spiralling phase space trajectories deep in the coexistence regime for $\lambda \gg \lambda_c$ that reflect damped oscillatory kinetics, yet with manifest fluctuations near the focal center that are apparent as the broad "blob" in the (green) curve for $\lambda = 0.25$ in Fig. 2(b). Intriguingly, these fluctuations become more prominent with increasing distance from the active-to-absorbing state transition point.[21,26,52] For very large predation rates, even sizeable systems become vulnerable to total population extinction.[41,55] While any accessible absorbing state constitutes the ultimately only stable configuration in a finite system, the typical extinction times scale exponentially with the system size L^d, whence one typically observes species coexistence instead for sufficiently large L.[56,57]

Following the time evolution of individual simulation runs (Fig. 3) provides enlightening information on the origin of the erratic population oscillations and measured pertinent correlations in stochastic lattice Lotka–

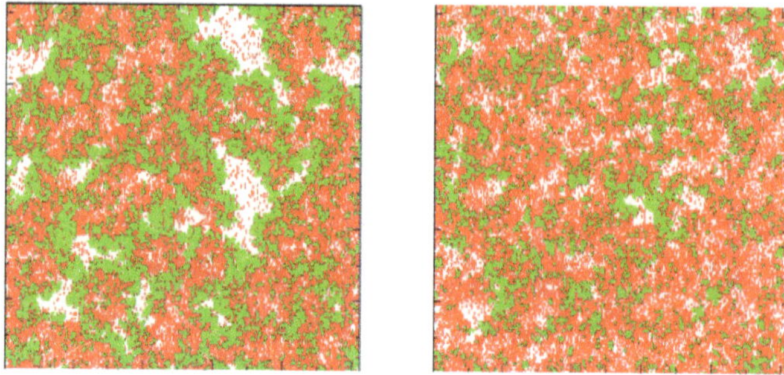

Figure 3.   Monte Carlo simulation snapshots of a stochastic Lotka–Volterra model on a 256×256 square lattice subject to periodic boundary conditions and site exclusion. Predators (red) and prey (green) are initially randomly distributed; white: empty spaces. Pictured are the configurations at 500 (left) and 1000 MCS (right). [Figures reproduced with permission from Ref. [72], copyright (2016) by IOP Publ.]

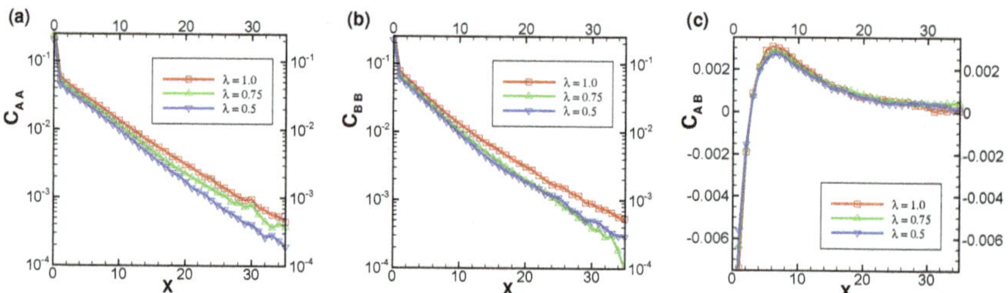

Figure 4.   Equal-time (cross-)correlation functions (a) $C_{AA}(x)$, (b) $C_{BB}(x)$, and (c) $C_{AB}(x)$, inferred from Monte Carlo simulations of the site-unrestricted stochastic Lotka–Volterra model on a 1024×1024 square lattice in the (quasi-)stationary regime, with rates $\mu = \sigma = 0.1$, and different predation rates $\lambda = 0.5$ (blue), 0.75 (green), 1.0 (red). [Figures reproduced with permission from Ref. [53], copyright (2007) by IOP Publ.]

Volterra systems.[58] For large predation rates λ, deep in the two-species coexistence region, predators may initially devour almost all the prey present in the system. Subsequently, in the absence of food, their number decays rapidly with rate μ and the lattice may move close to total population extinction. Yet through the branching processes at rate σ, surviving prey B serve as randomly placed sources for invasion fronts that spread into empty space; these are then effectively "followed" by spared predators A, leaving behind voids that may in turn be refilled by prey individuals.[10,21,26,52–54] As depicted in the simulation snapshots in Fig. 3, the underlying stochasticity thereby generates intriguing persistent spatio-temporal structures that may be viewed as fluctuation-driven Turing patterns.[59,60] As these noise-stabilized evasion-pursuit waves traverse the system and interact, they generate stochastic population oscillations for both species. Interestingly, although correlations and cross-correlations remain short-ranged, with correlation lengths typically spanning just a few lattice spacings (Fig. 4), the associated local population fluctuations are quite large.[21,26,53,54,61] Bigger systems may however accommodate more out-of-phase activity fronts; consequently the global particle number oscillations decrease in amplitude.[21]

On one-dimensional lattices with multiple particles permitted on each site, the ballistically propagating evasion-pursuit fronts periodically wipe out large spatial regions;[53,54] if severe site restrictions are enforced, one instead observes segregation into prey and predator domains, which slowly coalesce and coarsen.[21] In spatial dimensions $d \geq 4$, the spontaneously generated spatio-temporal structures are not visible anymore, and the system's dynamics essentially attains mean-field character. This is in fact characteristic of two-species binary reactions, exemplified by simple pair annihilation $A + B \to \emptyset$, which is known to display species segregation and hence spontaneous cluster formation only in low dimensions $d \leq d_s = 4$.[10,11,25,62–65] In stark contrast to the Lotka–Volterra mean-field rate equations, the corresponding stochastic spatially extended system turns out remarkably robust with respect to model modifications. For example, one may consider splitting the original binary reaction $A + B \to A + A$ perhaps more realistically into independent predation $A + B \to A$ and predator reproduction $A + B \to A + A + B$ processes, both requiring the presence of a prey individual adjacent to the predator. Whereas the mean-field scenario for the ensuing more complicated reaction scheme differs substantially from Eqs. (10), lattice Monte Carlo simulations for the altered stochastic reactions in fact yield the same qualitative behavior for both model variants.[52]

3.3. *Renormalized reaction rates and oscillation parameters*

Computing the Fourier transforms $\langle n_{A/B}(f) \rangle = \int \langle n_{A/B}(t) \rangle \, e^{2\pi i f t} \, dt$ of the population time tracks, the characteristic oscillation frequency may be extracted from the Fourier peak intensity, and the attenuation coefficient from its width. Results are displayed in Fig. 5(a), showing that the oscillation frequencies in the stochastic lattice Lotka–Volterra model are substantially reduced (by about 1/3) relative to the linearized mean-field prediction. This indicates strong renormalizations due to fluctuations, but the data

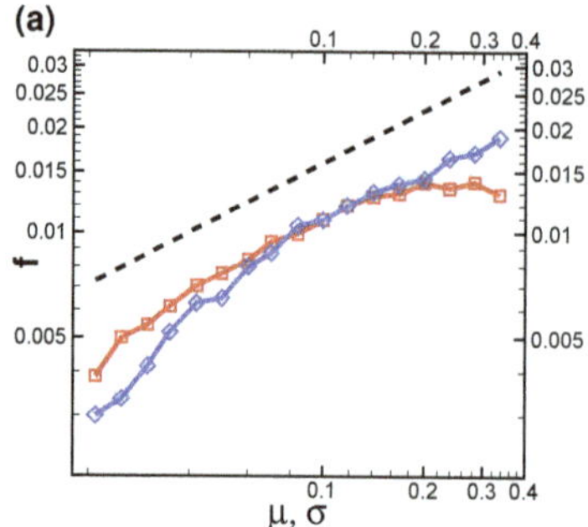

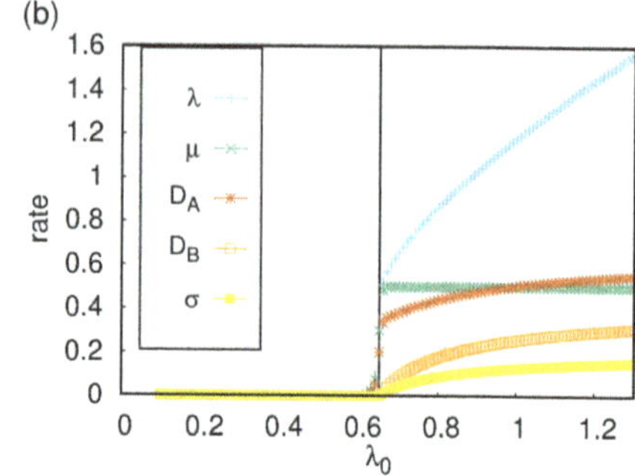

Figure 5. (a) Measured dependence of the characteristic peak frequency in the population number Fourier transforms $|\langle n_{A/B}(f)\rangle|$ on the rates σ (red squares) and μ (blue diamonds), with $\lambda = 1$ and the respective other rate held fixed at the value 0.1, obtained from simulation data for a stochastic Lotka–Volterra system on 1024×1024 square lattices without site restrictions up to time $t = 20{,}000$ MC; for comparison, the dashed black line shows the (linear) mean-field oscillation frequency $f_0 = \sqrt{\mu\sigma}/2\pi$. (b) Macroscopic reaction rates μ, σ, coupling λ, and diffusivities $D_{A/B}$ for the stochastic Lotka–Volterra predator-prey model on a square lattice with length $L = 150$, initial densities $a(0) = b(0) = 0.5$, as function of the microscopic predation rate λ_0, with fixed $\mu_0 = \sigma_0 = 0.5$, and $K = 1$; the vertical black line indicates the predator extinction / prey fixation threshold λ_c. [Figure (a) reproduced with permission from Ref. [53], copyright (2007) by IOP Publ.; (b) adapted from Ref. [39].]

still approximately follow a square-root dependence on the rates μ and σ.[53] A recent study systematically explores the deviations of the measured coarse-grained reaction rates from the "microscopic" parameters for the computer simulation algorithm.[39] The dependencies of the "macroscopic" predation coupling λ, predator death rate μ and prey birth rate σ, as well as both species' diffusivites $D_{A/B}$ on the input reactivity λ_0 are plotted in Fig. 5(b). As one should expect, the linear stochastic death processes occur independent of any spatial correlations, and hence μ remains unaltered until the predator extinction / prey fixation absorbing state transition at λ_c is reached, beyond which of course all reactions cease. As the prey begin to fill the site-restricted square lattice with imposed local carrying capacity $K = 1$, prey proliferation and hopping become diminished and tend to zero as $\lambda_0 \to \lambda_c$, with constant ratio σ/D_B. In contrast, the predators' effective reaction rates jump discontinuously at λ_c.[39]

The Doi–Peliti formalism maps the stochastic master equation kinetics to a non-Hermitean bosonic many-particle Hamiltonian and ultimately employs coherent states, *i.e.*, the complex eigenvalues of the annihilation operators, to construct a continuum field theory representation.[10,19,30–33] For the stochastic Lotka–Volterra predator-prey model with local site restrictions, the resulting action becomes[10,34]

$$S[\tilde{a}, \tilde{b}; a, b] = \int d^d x \int dt \left[\tilde{a} \left(\frac{\partial}{\partial t} - D_A \nabla^2 + \mu \right) a + \tilde{b} \left(\frac{\partial}{\partial t} - D_B \nabla^2 - \sigma \right) b \right.$$
$$\left. - \sigma \, \tilde{b}^2 \, b + \frac{\sigma}{K} \left(1 + \tilde{b} \right) \tilde{b} \, b^2 - \lambda \left(1 + \tilde{a} \right) \left(\tilde{a} - \tilde{b} \right) a \, b \right], \quad (14)$$

where the complex fields $a(x,t)$ and $b(x,t)$ originate from the coherent states, while the associated "response" fields are associated with their shifted adjoints, $\tilde{a} = \hat{a} - 1$, and similarly for $\tilde{b}$. The "classical" field equations associated with the action $\delta S/\delta a = 0 = \delta S/\delta b$ are solved by $\tilde{a} = 0 = \tilde{b}$, whereupon $\delta S/\delta \tilde{a} = 0 = \delta S/\delta \tilde{b}$ recovers the reaction-diffusion equations (13). Indeed, generally $\langle a(t) \rangle = \langle n_A(t) \rangle$ and $\langle b(t) \rangle = \langle n_B(t) \rangle$, but similar simple correspondences do not hold beyond the mean particle numbers.[10,19]

If one interprets the action (14) as the Janssen–De Dominicis functional in a path integral representation of stochastic partial differential equations,[36–38] it becomes equivalent to the coupled Langevin equations

$$\frac{\partial \, a(x,t)}{\partial t} = \left(D_A \nabla^2 - \mu \right) a(x,t) + \lambda \, a(x,t) \, b(x,t) + \zeta(x,t) , \quad (15)$$

$$\frac{\partial \, b(x,t)}{\partial t} = \left(D_B \nabla^2 + \sigma \right) b(x,t) - \frac{\sigma}{K} \, b(x,t)^2 - \lambda \, a(x,t) \, b(x,t) + \eta(x,t) ,$$

albeit for complex fields a, b.[10,34] These in effect add (complex) Gaussian stochastic forcings with zero mean $\langle \zeta \rangle = 0 = \langle \eta \rangle$ to Eqs. (13), subject to the noise (cross-)correlations

$$\langle \zeta(x,t) \, \zeta(x',t') \rangle = 2\lambda \, a(x,t) \, b(x,t) \, \delta(x - x') \, \delta(t - t') ,$$
$$\langle \zeta(x,t) \, \eta(x',t') \rangle = -\lambda \, a(x,t) \, b(x,t) \, \delta(x - x') \, \delta(t - t') ,$$
$$\langle \eta(x,t) \, \eta(x',t') \rangle = 2\sigma \, b(x,t) \left[1 - b(x,t)/K \right] \delta(x - x') \, \delta(t - t') , \quad (16)$$

which entail multiplicative reaction noise terms that vanish as the particle densities approach zero, as is appropriate for the presence of a fully absorbing extinction state at $a = 0 = b$. Similar Langevin equations can be derived by means of a van Kampen system size expansion.[27,59]

In the predator-prey coexistence region, the mean population numbers $\langle a \rangle$ and $\langle b \rangle$ are finite. Consequently, the Langevin equations (15) describe spatially distributed nonlinear oscillators driven by additive white noise.

The ensuing random kicks will occasionally be synchronous with the system's resonance frequency, causing large deviations from the coexistence fixed point densities, followed by attenuated population oscillations. This resonant amplification mechanism of internal stochastic fluctuations ultimately generates the persistent spatio-temporal structures and associated random oscillatory kinetics in Lotka–Volterra and related models.[10,21,34,66]

For a more detailed analysis, one may proceed by expanding the action in terms of fluctuating fields, defined relative to their mean stationary values. Diagonalization of the ensuing bilinear (Gaussian) action for equal predator and prey diffusivities $D_{A/B} = D_0$ yields circularly polarized modes with dispersion relation $i\omega(q) = \pm i\omega_0 + \gamma_0 + D_0\, q^2$, where $\omega_0 = \sqrt{\mu\,\sigma\,(1 - \mu/\lambda\,K) - \gamma_0^2}$ is the "bare" oscillation frequency and $\gamma_0 = \mu\,\sigma/2\lambda\,K$ denotes the damping coefficient, see Eq. (12). In the absence of site occupation restrictions $K \to \infty$, $\gamma_0 \to 0$. Through a systematic perturbative calculation in terms of the effective expansion parameter $g = (\lambda/\omega_0)(\omega_0/D_0)^{d/2}$, one may evaluate fluctuation corrections to the oscillation frequency, diffusivity, and attenuation.[10,34] Although the parameter regime where perturbation theory is valid cannot easily be accessed in computer simulations, since the predation reactions would be too rare to generate decent statistics, the calculations confirm some of the pertinent numerical results: At least to first order in g, the characteristic oscillation frequency ω is shifted downward; this renormalization is particularly strong for $d \leq 2$ owing to the destructive interference of the two oppositely polarized eigenmodes which are almost massless due to the weak damping. Also, the leading fluctuation corrections are symmetric in the rates σ, μ and are enhanced as $\sigma \ll \mu$ or $\sigma \gg \mu$; these features are in accord with the Monte Carlo data shown in Fig. 5(a). The diffusivity D is shifted upwards by the fluctuations, resulting in faster propagation of the evasion-pursuit fronts, while the attenuation γ is diminished compared to mean-field theory. Both these renormalizations help to induce instabilities towards spontaneous pattern formation that occur for $\gamma < 0$ at wavenumbers $q < \sqrt{|\gamma|/D}$.

3.4. *Predator extinction: Critical dynamics and aging*

Generically, the critical behavior at continuous nonequilibrium phase transitions from active to inactive, absorbing states is expected to be captured by the directed percolation universality class, at least in the absence of quenched disorder and any coupling to other conserved fields.[10,11,17–20,22,25,67,68] Since this Janssen–Grassberger conjecture also applies to reactive particle systems that incorporate multiple interacting species,[69,70] the observed dynamic scaling properties in stochastic spatial Lotka–Volterra models at the predator extinction threshold were immediately linked to critical directed percolation.[21,41,46,47,49,52] Heuristically, when near fixation the prey almost fill the entire system, the implicit condition for the predation reaction that one of the scarce predators must locate a prey individual is essentially always satisfied, whence the Lotka–Volterra processes reduce to the single-species (predator) reactions $A \to \emptyset$ and $A \rightleftharpoons A + A$, where the pair coagulation reaction originates from the site occupation restrictions or finite local carrying capacity.[21] These stochastic death, birth, and pair coagulation processes generate anisotropic directed percolation clusters in $(d+1)$-dimensional space-time, depicted in Fig. 14(a) in Sec. 5.1 below. When they all terminate after a finite time t, the system is in the absorbing state; when they extend to $t \to \infty$, it resides in the active phase. More formally, for $b \approx K$ the Doi–Peliti action (14) can be reduced to Reggeon field theory, which governs and in fact defines the directed percolation universality class.[10,11,21,25,26,34,67,71]

As Fig. 6 demonstrates, numerically extracting the dynamic scaling exponents from simulation data near the stationary regime at the stochastic Lotka–Volterra model's predator extinction / prey fixation phase transition does not produce satisfactory results even for fairly large lattices.[72] Indeed, double-logarithmic plots of the predator density decay, asymptotically to follow the power law $\langle n_A(t) \rangle \sim t^{-\alpha}$ at $\lambda = \lambda_c$, allow a precise determination of the critical point location (here at $\lambda_c = 0.0416$), independent of the lattice initialization. Yet a careful examination of the associated local slope or (time-dependent) effective decay exponent $\alpha_{\text{eff}}(t)$, displayed in the inset of Fig. 6(a), indicates that the universal asymptotic regime has barely been reached in these data, resulting in $\alpha \approx 0.54$; the accepted critical directed percolation value in two dimensions is $\alpha \approx 0.4505$.[10,22] Measuring the dynamic exponent $z\nu$ characterizing critical slowing down as obtained from analyzing the system's relaxation times $t_c \sim |\lambda - \lambda_c|^{-z\nu}$ following quenches near λ_c, depicted in Fig. 6(b), gives the satisfactory value $z\nu \approx 1.3$,[72] in accord with the literature exponents $z \approx 1.766$ and $\nu \approx 0.733$.[10,22]

Strong finite-size corrections may be avoided through seed dynamical Monte Carlo simulations, where initially a single active site, *i.e.*, a predator A, is placed in the lattice, with the remainder occupied by prey B.[17,18] At the extinction critical point, the ensuing predator survival probability and density in this seed initial-slip regime should scale as $P_A(t) \sim t^\delta$ and $\langle n_A(t) \rangle \sim t^\theta$, where for directed percolation $\delta = \alpha = \beta/z\nu$ and

                    U. C. Täuber

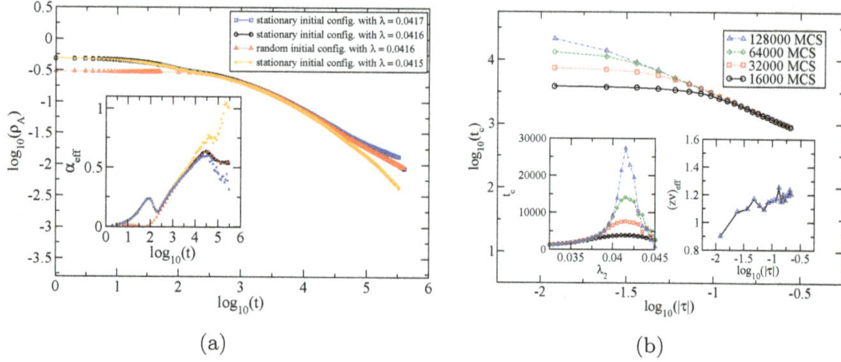

Figure 6.   (a) Predator decay (double-logarithmic plots) for a stochastic Lotka–Volterra model on a 1024×1024 square lattice with $\mu = 0.025$ and $\sigma = 1$ at the prey fixation threshold $\lambda_c = 0.0416$, for quasi-stationary (top, black) and random (lower curve, red dotted) initial configurations (data averaged over 2000 independent simulation runs). For comparison, the density decay data are also plotted for $\lambda = 0.0417$ (blue, active coexistence phase) and $\lambda = 0.0415$ (orange, predator extinction regime). The inset shows the effective decay exponent $\alpha_{\mathrm{eff}}(t)$, approaching $\alpha = 0.54 \pm 0.007$ at large t. (b) Characteristic relaxation time t_c measured near the critical point. Left inset: $t_c(\lambda_2)$ after the system is quenched from a quasi-steady state at $\lambda_1 = 0.25$ to much smaller values λ_2 near $\lambda_c = 0.0416$, where the different graphs indicate t_c when 128,000, 64,000, 32,000, and 16,000 MCS elapsed after the quench (data averaged over 500 runs). The main panel depicts these same data in double-logarithmic form; the graphs collapse for $|\tau| = |(\lambda_2/\lambda_c) - 1| > 0.1$, resulting in $z\nu = 1.208 \pm 0.167$. Right inset: associated effective exponent $(z\nu)_{\mathrm{eff}}(\tau)$, approaching $z\nu \approx 1.3$ as $|\tau| \to 0$. [Figures reproduced with permission from Ref. [72], copyright (2016) by IOP Publ.]

$\theta = (2 - \eta)/z = (d/z) - 2\alpha,$[10,20] with $\theta \approx 0.2295$ in $d = 2$ dimensions.[10,22] Numerical analysis of the Monte Carlo simulation data displayed in Fig. 7 for a stochastic Lotka–Volterra model on a 512×512 square lattice with site restrictions yields the accurate critical exponents $\delta \approx 0.451$ and $\theta \approx 0.230$.[21]

Accessing critical dynamical scaling exponents through the universal short-time "physical aging" regime is another convenient means to avoid running costly simulations for very large system sizes L that are required to meaningfully reach long times $t \sim L^z$.[10,73–76] To this end, one may for example suddenly switch the predation rate globally from λ_1 to a different value λ_2.[26,72] Outside the linear response regime, two-point (auto-) correlation functions $C(t,s) = \langle n(x,t)\, n(x,s) \rangle - \langle n(t) \rangle \langle n(s) \rangle$ will then depend on both involved times, often termed "waiting time" s (after the initial system preparation) and "observation time" $t > s$. Only if the system's

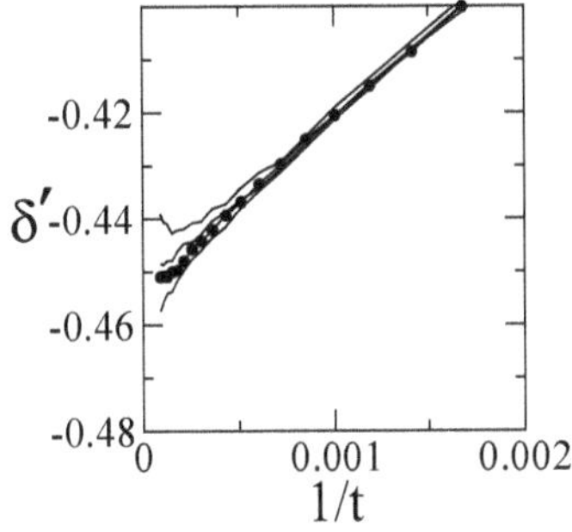 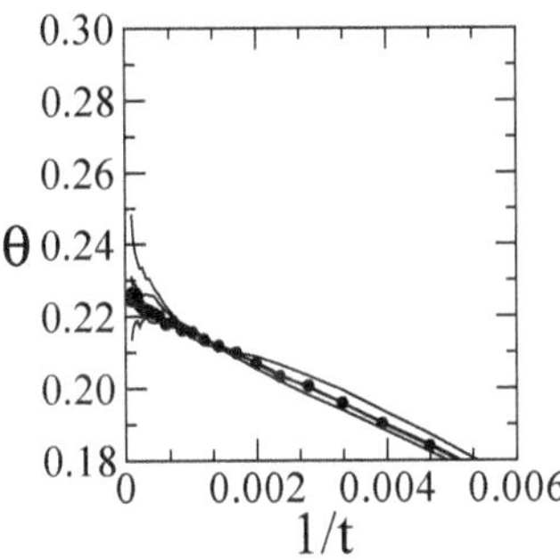

Figure 7. Seed dynamical Monte Carlo simulations to estimate the predator extinction threshold λ_c and critical exponents δ and ν for a stochastic site-restricted ($K = 1$) Lotka–Volterra model on a 512×512 square lattice with $\mu = 0.25$, $\sigma = 1$, $\lambda = 0.1690, 0.1689, 0.1688, 0.1687$ (top to bottom). The data are averaged over $3 \cdot 10^6$ independent runs with duration 10^5 MCS. The effective critical initial slip exponents $\delta_{\text{eff}}(t)$ for the predators' survival probability $P_A(t)$ (left) and $\theta_{\text{eff}}(t)$ for their mean particle number $\langle n_A(t) \rangle$ (right) are plotted vs. $1/t$, giving $\lambda_c \approx 0.1688$. Extrapolation to $t \to \infty$ yields the estimates $\delta \approx 0.451$ and $\theta \approx 0.230$. [Figures reproduced with permission from Ref. [21], copyright (2007) by Springer Science+Business Media, Inc.]

intrinsic relaxation processes are fast, typically exponential with characteristic time t_c, the preparation configuration prior to the quench is quickly forgotten, and time translation invariance recovered, as seen in the inset of Fig. 8(a), where both λ_1, λ_2 pertain to the coexistence phase.

In contrast, near a continuous phase transition, the relaxation time diverges according to a power law. Consequently, for $t \gg s$ the two-time order parameter autocorrelation function obeys the scaling form[10,20,25,73,75,76]

$$C(t, s) = s^{-b} f_c(t/s) , \quad f_c(y) \sim y^{-\Lambda_c/z} , \tag{17}$$

which defines the aging scaling exponent b and the autocorrelation exponent Λ_c.[76] For critical directed percolation the scaling relations $b = 2\alpha$ and $\Lambda_c/z = 1 + \alpha + (d/z)$ hold.[10,20,72] The numerical simulation data on a 1024×1024 square lattice with stochastic Lotka–Volterra kinetics demonstrates that time translation invariance is broken in quenches to the predator extinction / prey fixation threshold λ_c, c.f. Fig. 8(a), main panel. Instead, as demonstrated in Fig. 8(b), for sufficiently large waiting times $s \geq 1000$ MCS the aging scaling form (17) applies. The best data scaling collapse is obtained for $b \approx 0.88$, and the resulting slope of the master curve ultimately yields $\Lambda_c/z \approx 2.37$, in decent agreement with the established two-dimensional critical directed percolation exponents.[26,72] Intriguingly, the onset of shear turbulence in pipe flow displays spatio-temporal phe-

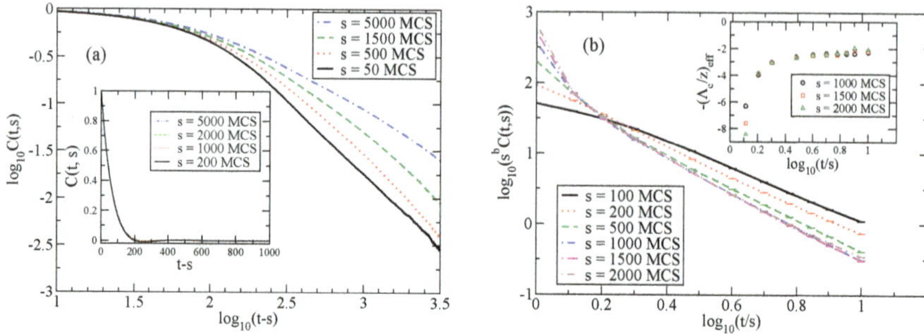

Figure 8.   (a) Double-logarithmic plot of the predator density autocorrelation function $C(t, s)$ vs. time difference $t - s$ for various waiting times $s = 50, 500, 1500, 5000$ MCS (left to right) at the predator extinction critical point $\lambda_c = 0.0416$ for the same Lotka–Volterra system as in Fig. 6 (data averaged over 100 independent simulation runs for each value of s). The inset shows that $C(t, s)$ decays exponentially for $\lambda_2 = 0.125$, *i.e.*, a quench within the coexistence phase; here time translation invariance holds (data averaged over 400 simulation runs for each $s = 5000, 2000, 1000, 200$ MCS, top to bottom). (b) Simple aging dynamical scaling analysis: $s^b\, C(t, s)$ is plotted against the time ratio t/s, with 1000 independent simulation runs for each s. The straight-slope section of the curves with large $s \geq 1000$ MCS yields $\Lambda_c/z = 2.37 \pm 0.19$; the aging scaling exponent is $b = 0.879 \pm 0.005$. The inset displays the local effective exponent $-(\Lambda_c/z)_{\text{eff}}(t)$. [Figures reproduced with permission from Ref. [72], copyright (2016) by IOP Publ.]

nomena akin to predator-prey kinetics; hence its threshold properties are captured by the directed percolation universality class as well.[77] Indeed, the associated critical exponents have been most accurately measured experimentally near the transition between two different turbulent states of electrohydrodynamic convection in thin nematic liquid crystals.[78,79]

4. Biologically motivated model variants and extensions

Naturally, a large variety of possible modifications and extensions can be envisioned for the simple Lotka–Volterra model for prey-predator competition and coexistence. Indeed, within the realm of coupled nonlinear mean-field rate equations, there exists a multitude of such generalizations studied in a vast literature encompassing different fields. Biological applications pertain to realizations in ecology, addressing predator-prey and host-parasite systems on all scales reaching from terrestrial mammals and fish populations to the microbiology of competing bacteria and oceanic plankton.

Stochastic spatially extended model implementations have been less frequently employed, but are becoming increasingly appreciated.[16] The complexity of multi-species ecological systems subject to many ill-defined external influences may be avoided in synthetic laboratory settings that are subject to controlled initial and environmental conditions. In that context, simple models in the spirit of the paradigmatic Lotka–Volterra system definitely constitute a powerful tool to both qualitatively and quantitatively describe experimental data; one example captures spatial range expansion of competing bacterial colonies, one of which may acquire resistance against toxins produced by the other species.[80]

On the theoretical front, the stability of multi-species systems and hence emergence of biodiversity remains an ecological puzzle, since within a mean-field framework, only the strongest predators or most evasive prey should survive direct competition. This scenario is however not confirmed in the corresponding stochastic model setups: For two distinct populations preyed upon by a single predator species, intrinsic demographic fluctuations are observed to prevent extinction of the weaker prey type.[81] Likewise, three-species coexistence in a system of two different predator groups competing for a single prey population may likewise be stabilized by stochasticity, especially when subject to evolutionary optimization.[82] In the following, just three biologically motivated Lotka–Volterra model variants are described in more detail, namely the system's response to periodic (seasonal) switches of the carrying capacity;[39] the effect of quenched random spatial disorder in the nonlinear predation rates mimicking environmental variability;[61] and the ensuing evolutionary dynamics upon introducing (in addition) individual demographic variability in the predation efficacy, whose distributions in both predator and prey species adjust themselves through the Darwinian mechanisms of competition and inheritance with random mutations.[26,54,83] More substantial variations of the Lotka–Volterra predation paradigm with prominent biological relevance are briefly discussed in Secs. 5 and 6 on stochastic models for epidemic outbreaks and cyclic competition.

4.1. *Periodically varying carrying capacity*

Seasonal variations in resource availabilty, especially for the prey population, may be encoded in a periodically switching carrying capacity K; similarly, bacterial populations thriving on agar solutions in Petri dishes may be supplied with nutrients at regular time intervals. In a stochastic Lotka–Volterra lattice model, one may directly implement a "rectangular time signal" for $K(t)$, switching from K_- to $K_+ > K_-$ and back at full

                       U. C. Täuber

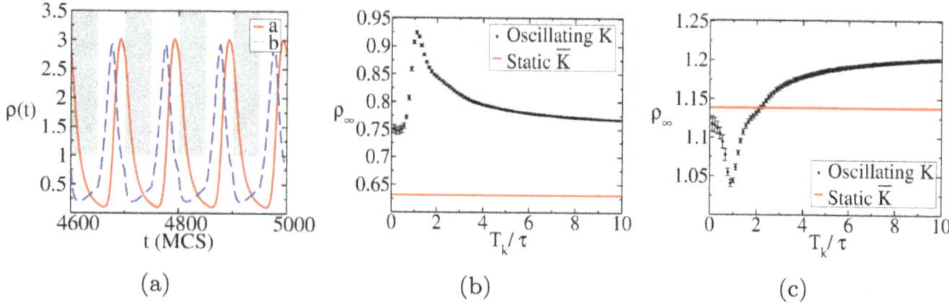

Figure 9.   (a) Predator (full red) and prey (dashed blue) populations (averaged over 50 realizations) for a 256×256 square lattice with $\mu = \sigma = \lambda = 0.1$, $K_- = 1$, $K_+ = 10$, and $T_k = 100$; the shaded gray indicate the areas excluded by the switching carrying capacity $K(t)$. The critical predation rates associated with fixed carrying capacities are $\lambda_c(K = 1) = 0.26$ and $\lambda_c(K = 10) = 0.01$. Long-time (b) predator and (c) prey populations averaged over six periods T_k plotted vs. T_k/τ, where τ denotes the intrinsic oscillation period of the equivalent static system with $K = \bar{K}$; $L = 256$, $\mu = \sigma = \lambda = 0.1$, and $K_- = 2$, $K_+ = 6$ for the oscillating environment (black crosses), while $K = \bar{K} = 3$ for the static environment (red horizontal lines). [Figures reproduced with permission from Ref. [55], copyright (2023) by The American Physical Society.]

period T_k, in the simplest symmetric scenario remaining at $K_\pm$ for the duration $T_k/2$. If the entire carrying capacity range pertains to the species coexistence phase, only a minor amplification of the periodic population oscillations is observed. Conversely, predator extinction would of course ensue if $\lambda < \lambda_c(K_\pm)$. However, as shown in Fig. 9(a), the resulting time tracks for both predators and prey display large-amplitude periodicity if the carrying capacity oscillates between the predator extinction and species coexistence phases, *i.e.*, in a static system with carrying capacity K_-, the A population would die out, but switching to K_+ above the critical threshold for a sufficiently long time allows the predators to recover.[55] For very fast switching, where T_k is much smaller than the characteristic intrinsic population oscillation period τ, the system becomes equivalent to an effective static model. Since in mean-field theory the stationary predator and prey populations $\sim 1/K$, their densities in the rapidly oscillating system are determined by the harmonic mean $\bar{K} = 2K_- K_+/(K_- + K_+)$. In the opposite extreme limit of very large $T_k \gg \tau$, both populations (in the coexistence phase) essentially merely switch between their stationary values, which is described by an equivalent static model with a rate-dependent carrying capacity K^* that reduces to $\bar{K}$ if both $K_\pm \gg 1$.[55]

Monte Carlo simulation results for stochastic lattice Lotka–Volterra

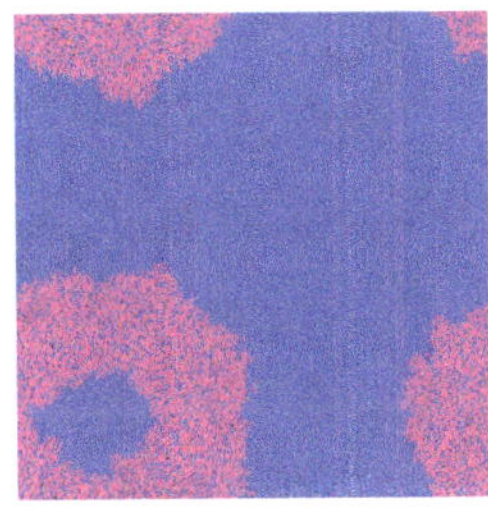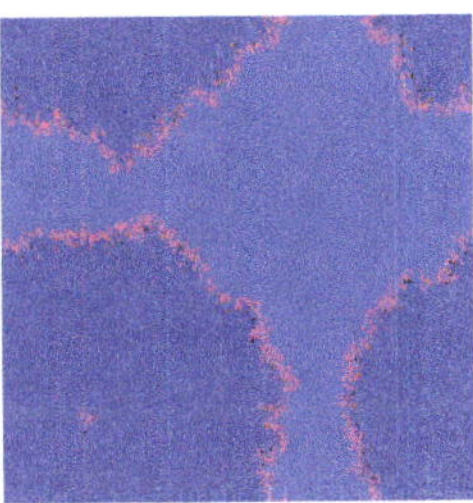

Figure 10. Simulation snapshots for a 256×256 square lattice with $\mu = \sigma = 0.5$, $\lambda = 0.1$, $K_- = 1$, $K_+ = 10$, and $T_k = 30$ MCS at $t = 39$ (left), 56 (middle), 72 MCS (right); red / blue pixels indicate predators / prey, where the brightness represents the local density, pink pixels pertain to sites occupied by both species; black: empty sites. The system is initialized with $K(t = 0) = K_- = 1$. [Figures reproduced with permission from Ref. [55], copyright (2023) by The American Physical Society.]

models with periodically switching $K(t)$ confirm that the prey population is indeed generally quite well described by these effective static limits, within $\approx 10\%$ for the measured mean prey density, as seen in Fig. 9(c). Yet the mean predator population in Fig. 9(b) displays manifest quantitative deviations from the mean-field predictions, but of course approaches (slightly different) constant limits in both fast- and slow-switching regimes. Near $T_k \approx \tau$, intriguing resonances are visible. Related simulation snapshots are depicted in Fig. 10 for a switching period $T_k = 30$ MCS, close to resonance.[55] In this simulation run, by $t = 39$ MCS (left) only a single predator patch has survived, subsequently serving as the source for a spreading population front. At $t = 56$ MCS (middle) this wave expands in- and outward, until by $t = 72$ MCS (right) it returns to the source center; just then the system is switched back to the low carrying capacity K_-, whence the prey in the front interior cannot reproduce anymore. In its later temporal evolution, the system's kinetics continues to be dictated by the single population oscillation source generated by the sole predator patch surviving early on.

4.2. *Fitness enhancement through environmental variability*

Natural environments are of course not homogeneous, and subject to spatially varying resource availability, and specific to predator-prey interactions, habitat features that may locally enhance or suppress the chance of predation events. Such scenarios can be modeled through a spatially varying carrying capacity K, or site-dependent reaction rates, either smoothly varying or distinct in different regions or "patches". A perhaps extreme sit-

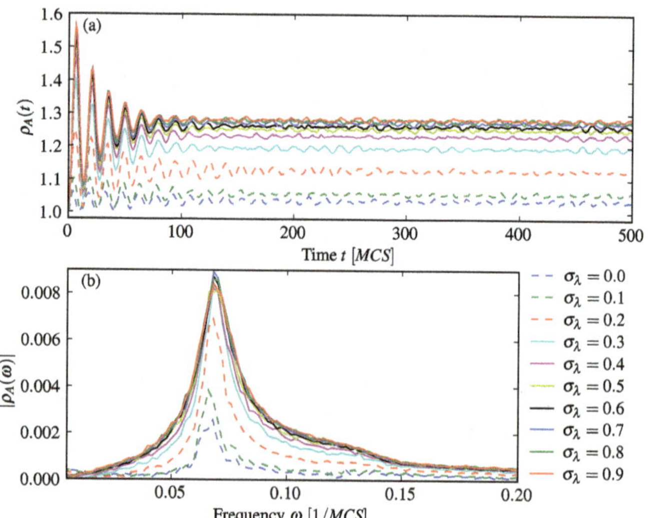

Figure 11.   (a) Predator population time evolution in a site-disordered Lotka–Volterra model on a 512×512 square lattice with $\mu = \sigma = 0.5$, mean predation rate $\bar{\lambda} = 0.5$, for different variances σ_λ as indicated; data averaged over 50 independent simulation runs. (b) Fourier transform of the time traces. [Figure reproduced with permission from Ref. [61], copyright (2008) by The American Physical Society.]

uation is to assign randomized rates to each discrete lattice site, picked for example from a Gaussian distribution, truncated for the reaction probabilities to the interval $[0, 1]$, and centered at a prescribed mean value with an assigned variance that indicates the "strength" for this quenched disorder.

As one would anticipate, varying rates μ, σ for the linear predator death and prey birth processes yield negligible changes to the mean populations, as random shifts just average out. That is not true if the predation rates are drawn from truncated Gaussian distributions with average $\bar{\lambda}$ and variance σ_λ, since (in the mean-field approximation) both stationary coexistence populations scale inversely with λ. Consequently, sites with low predation rates turn out favorable for, perhaps counterintuitively, both predator and prey species.[26,61]   Indeed, as shown in Fig. 11(a), Monte Carlo simulations for two-dimensional Lotka–Volterra systems with quenched predation rate site disorder yield a $\sim 25\%$ "fitness" enhancement in the stationary densities of predators and prey as the width σ_λ is increased towards an asymptotically flat distribution. Increasing disorder strength leads to the expected broadening of the associated Fourier peaks displayed in Fig. 11(b), which implies shorter relaxation times to reach the (quasi-)stationary configurations. Localization of the largest-amplitude local population oscillators near the favored sites with small λ values additionally reduces the (cross-)correlation lengths by $\sim 1/3$ relative to the homogeneous system. Consequently more population centers can be accommodated by a system of fixed size, which explains the resulting enhancement in both predator and prey numbers.[26,61]

4.3. *Demographic variability, hereditary trait optimization*

In contrast to physical and chemical reaction processes that involve fundamentally identical atoms, molecules, or similar microscopic constituents, individuals in a biological population are usually endowed with different propensities such as fecundity and efficacy in pursuing prey or evading predators. We may thus affix each reacting particle with, for example, a predation efficacy η. If a predator A and prey B encounter each other, we can for simplicity choose the resulting predation probability for this specific pair as the arithmetic mean $\lambda = (\eta_A + \eta_B)/2$. These individual "hunting" or "evasion" efficacies can be turned into random variables drawn from, *e.g.*, a Gaussian distribution with width w_S truncated to the interval $[0, 1]$: Powerful predators are represented by $\eta_A \approx 1$, whereas evasion-proficient prey are characterized by $\eta_B \approx 0$. In this manner, genetic make-ups and behavioral features are subsumed into one "mesoscopic" phenotypical trait.[26,54,83]

This setup allows us to invoke rudimentary Darwinian evolutionary dynamics by allowing offspring to inherit their specific efficacy η_O from their parent's individual η_P value, with a random mutation "error". As Fig. 12 illustrates, in the (a) prey reproduction $P \to P + O$ at fixed rate σ and (b) predation process $P + B \to P + O$ at variable rate $\lambda = (\eta_P + \eta_B)/2$, η_P is set at the center of truncated Gaussian distribution with standard deviation (prior to truncation) w_P, from which η_O is selected. The parameter w_P describes the mutation strength; for $w_P = 0$ the offspring assumes precisely its parent's efficacy η_P. Since surviving prey will likely pass their small η values to their progeny, while fertile predators will typically be endowed by large predation efficacies, the trait distributions for the two distinct populations will evolve in opposite directions.

Indeed, in Monte Carlo simulations of this model on a two-dimensional lattice both species' trait distributions settle, within a few hundred MCS (roughly equivalent to "generations"), to their stationary distributions shown in Fig. 13(a) and (b), respectively for a moderate mutation distri-

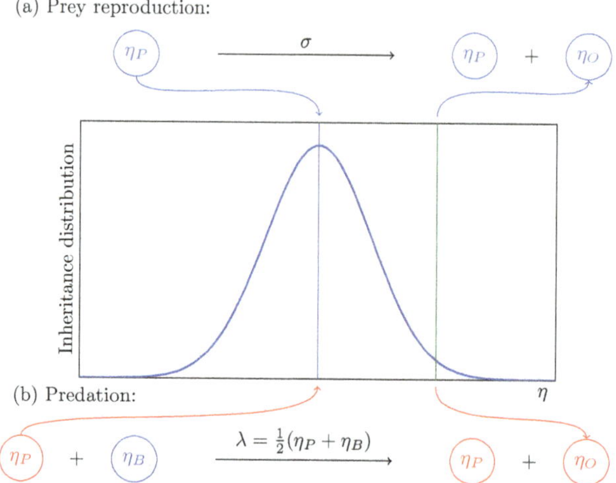

Figure 12. (a) During prey reproduction, a parent particle gives birth with rate σ. The parent's predation efficacy η_P is set as the mean of a Gaussian distribution, truncated to the interval $[0; 1]$, from which the offspring's efficacy η_O is drawn. (b) During predation, a predator consumes a prey individual with rate λ, a function of the participating particles' predation efficacies. An offspring predator is created and its efficacy η_O is determined via the same mechanism as for prey reproduction. [Figure reproduced with permission from Ref. [54], copyright (2013) by IOP Publ.]

bution width $w_P = 0.1$ and a flat distribution with $w_P \to \infty$. In the former scenario, the widths of the ensuing steady-state efficacy distributions for either species are comparable with w_P. In contrast with (linear) genetic drift models, this nonlinear evolutionary dynamics does not terminate in fixation of the prey at $\eta = 0$ and predators at $\eta = 1$; rather, demographic variabilities are sustained in both populations.[26,54,83] The dashed graphs indicate the numerical solutions of an effective mean-field model, wherein the traits η are discretized and binned to define multiple "quasi-species" that are subject to the standard Lotka–Volterra rate equations. In the extreme case $w_P \to \infty$, the predators' selection bias disappears, and the stationary prey distribution $\sim 1/(1 + 2\eta)$ for the quasi-species model can be evaluated analytically, and decently captures the simulation data.

Finally, the combined effects of environmental and demographic variability in the predation processes can be studied.[26,54,83] To this end, random reaction reaction rates λ_E attached to the lattice sites as well as predation efficacies $\eta_{A/B}$ relating to individual particles are implemented, each drawn from truncated Gaussian distributions with widths w_E and w_D, re-

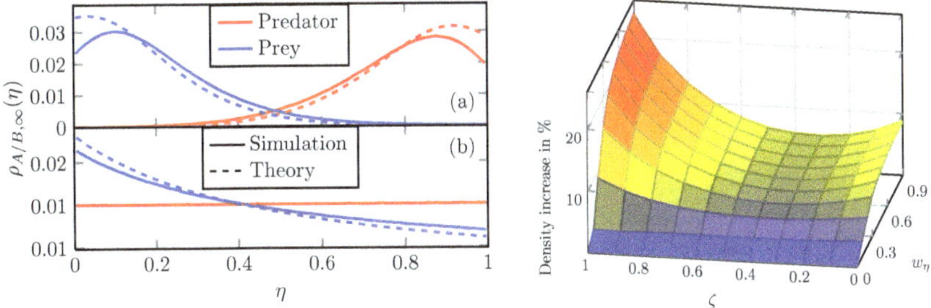

Figure 13. Left panel: Predator (red) and prey (blue) populations in the (quasi-)steady state of a stochastic Lotka–Volterra model (no site occupation restrictions) on a 128×128 square lattice with $\mu = \sigma = 0.5$ and variable, hereditary predation effiacies η affixed to all individuals, initially set to 0.5, as functions of η (data averaged over 10^4 Monte Carlo runs) for initial truncated Gaussian standard deviations $w_P = 0.1$ (a) and $w_P \to \infty$ (b). For finite w_P, the predator / prey distributions quickly evolve towards high / low predation efficacies. In the special flat distribution case, the predators experience no selection bias. The predictions from an effective stochastic multiple quasi-species mean-field model are depicted as dashed curves. Right panel: The stationary predator density features enhancement for all values of the spatial variability influence ζ as function of equal variabilities $w_W = w_D = w_\eta$ relative to uniform systems with $w_\eta = 0$; there appears a shallow minimum near $\zeta \approx 0.3$. [Figures reproduced with permission from Ref. [83], copyright (2013) by The American Physical Society.]

spectively. In order to tune the relative influences of quenched spatial randomness and variabilities of the predator and prey populations, which are set to evolve in time according to the offspring hereditary selection with mutation chance described previously, we choose the linear interpolation $\lambda = \zeta \lambda_E + (1 - \zeta) \lambda_D$, where $\lambda_D = (\eta_A + \eta_B)/2$ as before, and $0 \le \zeta \le 1$. The ensuing Monte Carlo simulation results for the stationary predator density are displayed in the right panel of Fig. 13. For $\zeta = 1$, the $\sim 25\%$ fitness enhancement for exclusive site disorder described in Sec. 4.2 is recovered. In contrast, pure demographic variability $\zeta = 0$ increases the populations by less than 10%, indicating that the substantial evolutionary adjustments of both species' predation efficacy distributions almost cancel out in their competitive optimization leading to an almost neutral overall benefit. At $\zeta \approx 0.3$, an distinct minimum with even lower population enhancement is discernible; it can be understood in terms of straightforward error progression assuming statistical independence of environmental and demographical variability. The data displayed here demonstrate the dominance of environmental randomness. Yet in small systems the mean extinction time is increased more than fourfold in the presence of demographic variability

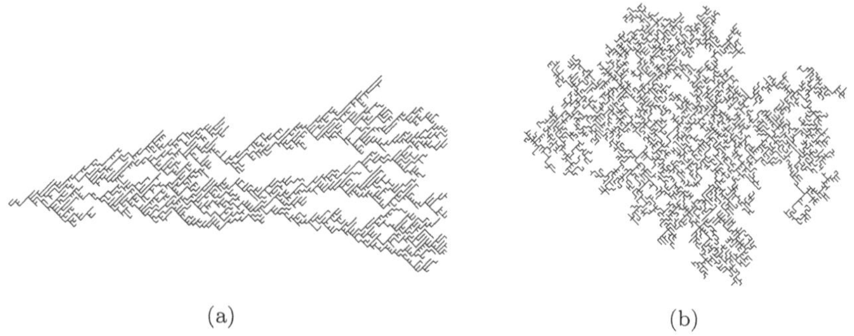

(a)   (b)

Figure 14. (a) Critical directed percolation cluster (time runs from left to right); (b) critical isotropic percolation cluster. [Figures reproduced with permission from Ref. [85], copyright (1994) by The American Physical Society.]

which thus manifestly provides a major evolutionary advantage.[54,83]

5. Stochastic spatial models for infectious disease spreading

The basic Lotka–Volterra predation motif also appears in the fundamental mathematical models for epidemic outbreaks and spreading, dating back to Kermack, McKendrick, and Walker's seminal work from 1927.[2,84] Instead of predators and prey or parasites and hosts, one considers a population comprised of a now fixed total number N of individuals that can be grouped into distinct states with respect to a contagious pathogen. To render these models more versatile and realistic, such "compartments" can be elaborated further by dividing them into additional distinct groups according to age or social structure, *etc.*, and by implementing spatially extended stochastic models on appropriate contact network architectures.

5.1. SIS *model and directed percolation*

In the simplest scenario, healthy susceptible individuals S become infected with rate r upon encountering sick or asymptomatic infectious agents I, following the binary (stochastic) reaction $S + I \xrightarrow{r} I + I$. The infected individuals may return to the susceptible group with recovery rate a, $I \xrightarrow{a} S$.[2,84] Note that these processes strictly conserve $N = n_S(t) + n_I(t)$; the latter transmutation may be regarded as number-conserving combination of the linear predator death and prey birth reactions of the Lotka–Volterra model, as is evident also in the corresponding mean-field rate equations

$$\frac{\partial \langle n_S(t) \rangle}{\partial t} = -r \, \langle n_S(t) \rangle \, \langle n_I(t) \rangle + a \, \langle n_I(t) \rangle \, ,$$

$$\frac{\partial \langle n_I(t) \rangle}{\partial t} = r \, \langle n_S(t) \rangle \, \langle n_I(t) \rangle - a \, \langle n_I(t) \rangle \, . \tag{18}$$

Near the epidemic threshold, there are few infected individuals $n_I \ll n_S$; the S species is abundant and may thus be eliminated (integrated out). Thus, upon implementing any population-limiting mechanism such as lattice site restrictions, this "susceptible-infected-susceptible" (SIS) model near criticality reduces to the elementary death, birth, and coagulation reactions that generate directed percolation clusters; an example is shown in Fig. 14(a).

A continuum description for "simple epidemic processes" that encompass the SIS model can be formulated in terms of a Langevin stochastic partial differential equation for the local density of infectious individuals $n(x,t)$ that spread diffusively and are subject to reactions which require the presence of "active" agents,[10,17–20,22,25,67]

$$\frac{\partial n(x,t)}{\partial t} = \left(D\nabla^2 - \mathcal{R}[n(x,t)] \right) n(x,t) + \eta(x,t) \, . \tag{19}$$

In the threshold vicinity, *i.e.*, near the continuous active-to-absorbing phase transition, the deterministic reaction kernel may be expanded according to $\mathcal{R}[n] \approx a + u\,n + \ldots$, which leads directly to the Fisher–Kolmogorov–Petrovsky–Piskunov equation.[2] The noise correlations too must obey the absorbing-state constraint and vanish as $n \to 0$. To leading order in the active density field, one arrives at the "square-root" multiplicative noise

$$\langle \eta(x,t) \, \eta(x',t') \rangle \approx v \, n(x,t) \, \delta(x - x') \, \delta(t - t') \, . \tag{20}$$

The Janssen–DeDominicis functional[10,36–38] associated with this Langevin kinetics (19), (20) is precisely the Reggeon field theory action.[10,19,20,71]

5.2. SIR *model and dynamic isotropic percolation*

After exposure to most infectious agents, individuals do not return to a susceptible state, but at least for some extended time period acquire immunity against renewed infection. Kermack, McKendrick, and Walker hence introduced a "recovered" compartment that in general incorporates both immune and deceased formerly infected populations; in this "susceptible-infected-recovered" (SIR) model the R species does not further partake in the system's dynamics. In a stochastic formulation, the associated reaction processes are $S + I \xrightarrow{r} I + I$ and $I \xrightarrow{a} R$. The total population number

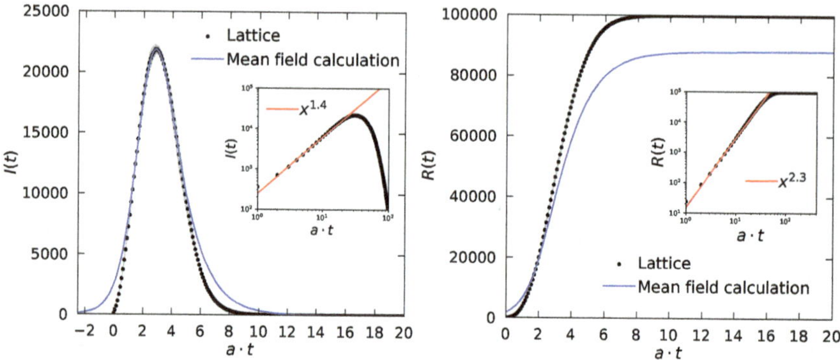

Figure 15.   Individual-based Monte Carlo simulations for a stochastic SIR model on a 448×448 square lattice with $N = 100{,}000$ individuals, initially $n_\mathrm{I}(0) = 100$, $n_\mathrm{R}(0) = 0$, infectivity $r = 1$, mean recovery time $1/a = 6.667$ days, giving $\mathcal{R}_0 = 2.4$ adequate for the original Covid-19 viral strain. The graphs show first of the infection curves from lattice simulations (averaged over 100 independent runs) to numerical integrations of the mean-field SIR rate equations. Left: Infectious population $\langle n_\mathrm{I}(t)\rangle$; right: recovered individuals $\langle n_\mathrm{R}(t)\rangle$. [Figure reproduced with permission from Ref. [86], copyright (2021) by Nature Research.]

$N = n_\mathrm{S}(t) + n_\mathrm{I}(t) + n_\mathrm{R}(t)$ remains fixed, but the infected fraction $n_\mathrm{I}(t)$ ultimately becomes depleted as it enters the R compartment. From the associated rate equations

$$\frac{\partial \langle n_\mathrm{S}(t)\rangle}{\partial t} = -r \, \langle n_\mathrm{S}(t)\rangle \, \langle n_\mathrm{I}(t)\rangle \, ,$$

$$\frac{\partial \langle n_\mathrm{I}(t)\rangle}{\partial t} = r \, \langle n_\mathrm{S}(t)\rangle \, \langle n_\mathrm{I}(t)\rangle - a \, \langle n_\mathrm{I}(t)\rangle \, ,$$

$$\frac{\partial \langle n_\mathrm{R}(t)\rangle}{\partial t} = a \, \langle n_\mathrm{I}(t)\rangle \, , \tag{21}$$

one obtains $d\langle n_\mathrm{S}\rangle/d\langle n_\mathrm{R}\rangle = -\langle n_\mathrm{S}\rangle \, r/a$, which with $\langle n_\mathrm{R}(0)\rangle = 0$ integrates to $\langle n_\mathrm{S}(\infty)\rangle = \langle n_\mathrm{S}(0)\rangle \, e^{-\langle n_\mathrm{R}(\infty)\rangle \, r/a} \geq \langle n_\mathrm{S}(0)\rangle \, e^{-N \, r/a} > 0$ and hence $\langle n_\mathrm{R}(\infty)\rangle = a \int_0^\infty \langle n_\mathrm{I}(t)\rangle \, dt = N - \langle n_\mathrm{S}(\infty)\rangle < N$, since $\langle n_\mathrm{I}(\infty)\rangle = 0$. Under the well-mixed mean-field assumption, $\langle n_\mathrm{I}(t)\rangle$ will initially increase and thus induce an epidemic outbreak if $\langle n_\mathrm{S}(0)\rangle > a/r$, or the (mean) "basic reproduction number" $\mathcal{R}_0 = \langle n_\mathrm{S}(0)\rangle \, r/a > 1$; otherwise $\langle n_\mathrm{I}(t)\rangle \to 0$ monotonically and the infection fizzles out.[2]

In spatially extended stochastic SIR models one observes distinct spreading infection fronts, similar to the activity waves in the Lotka–

Volterra predator-prey system. Fig. 15 compares the numerical solutions of the SIR rate equations for $\langle n_I(t)\rangle$ and $\langle n_R(t)\rangle$ with the corresponding mean populations extracted from individual-based Monte Carlo simulations for a stochastic model realization on a regular square lattice.[86] To this end, the effective "renormalized" rates (specifically the infectivity r) in the deterministic mean-field equations were adjusted to closely fit the peak intensity of the infection outbreak. Clear deviations between the rate equation predictions and the simulation data that properly incorporate fluctuations and correlations are discernible both at the epidemic's onset and at its late stages: Mean-field theory consistently overestimates $\langle n_I(t)\rangle$ (left panel), providing a too pessimistic outlook for controlling the disease outbreak such as enforcing regular testing protocols, quarantine of infected individuals, and isolation of their contacts.[87] The deterministic mean-field equations yield an initial exponential growth for the infected population at rate $a\,(\mathcal{R}_0 - 1)$; the simulation data are rather described by power laws for both $\langle n_I(t)\rangle$ and $\langle n_R(t)\rangle$, as shown in the insets in Fig. 15. Also, if the individuals are mobile (*e.g.*, via nearest-neighbor hopping), the entire susceptible population becomes infected, so ultimately $\langle n_S(\infty)\rangle = \langle n_I(\infty)\rangle = 0$ and $\langle n_R(\infty)\rangle = N$. The mean-field approximation then drastically underestimates $\langle n_R(\infty)\rangle$, by $\sim 15\%$ for the parameters chosen here (right panel).

The SIR model represents a wider class of "general epidemic processes" that entails the effective removal of (permanently) immune or deceased individuals, but also supposes that all previously infected agents I remain contagious to adjacent susceptibles S until they transition to the R class. This persistent infectivity introduces temporal memory; in the corresponding continuum representation, integrating out the infectious agent field again yields a Langevin equation of the form (19) with multiplicative noise (20), but with a reaction kernel $\mathcal{R}[n] \approx a + u \int^t n(x, t')dt' + \ldots$ that contains the accumulated "debris". This defines the dynamic isotropic percolation universality class.[10,17,18,20,22,25,88–90] Indeed, for $t \to \infty$ the associated Janssen–De Dominicis functional reduces to the stationary effective action that governs critical isotropic percolation clusters, see Fig. 14(b). Numerical simulations confirm these assertions even in the presence of lattice site dilution, representing randomly distributed immunized individuals.[91]

6. Cyclic dominance of three competing species

In ecological "food networks", there exist also cyclic competition motifs in addition to hierarchical food chain structures.[6,7,26,92] Such cyclically implemented dominance akin to the popular "rock-paper-scissors" children's game plays a prominent role in evolutionary game theory as well,[8,9] and may be experimentally realized in microbial colonies.[23,24,26] We will next briefly discuss some pertinent features of the two probably most frequently studied model variants for three species cyclically preying on each other.

6.1. *Rock-paper-scissors or cyclic Lotka–Volterra model*

The simplest rock-paper-scissors model just cyclically combines the Lotka–Volterra predation reactions $A + B \xrightarrow{\lambda_A} A + A$, *etc.*[8,9] Importantly, the total particle number $N = n_A(t) + n_B(t) + n_C(t)$ is strictly conserved. If we allow the three particle species to propagate via hopping to (or position swaps with) adjacent sites, the corresponding continuous density field obeys the diffusion equation. Similar to the two-species predator-prey system with infinite carrying capacity, the associated mean-field rate equations

$$\frac{\partial \langle n_A(t) \rangle}{\partial t} = \langle n_A(t) \rangle \left[\lambda_A \langle n_B(t) \rangle - \lambda_C \langle n_C(t) \rangle \right], \quad \text{(and cyclic)} \tag{22}$$

feature the coexistence fixed point $\langle n_A(\infty) \rangle = N \lambda_B / (\lambda_A + \lambda_B + \lambda_C)$, *etc.*, with linear stability matrix

$$\mathbf{L} = \frac{N}{\lambda_A + \lambda_B + \lambda_C} \begin{pmatrix} 0 & \lambda_A\,\lambda_B & -\lambda_B\,\lambda_C \\ -\lambda_A\,\lambda_C & 0 & \lambda_B\,\lambda_C \\ \lambda_A\,\lambda_C & -\lambda_A\,\lambda_B & 0 \end{pmatrix},$$

with eigenvalues 0 (for the conserved total particle number N) and $\pm i\omega_0$ describing neutral cycles and hence undamped population oscillations with frequency $\omega_0 = N\sqrt{\lambda_A\,\lambda_B\,\lambda_C / (\lambda_A + \lambda_B + \lambda_C)}$.

In a spatially extended stochastic cyclic Lotka–Volterra system, one might thus expect the emergence of activity waves akin to the two-species predator-prey model.

However, these are not actually observed; instead, such rock-paper-scissors lattice models are characterized by mildly fluctuating particle clusters of the same species,[93,94] see Fig. 16(a). Evidently the diffusive mode associated with the total particle number conservation is quite effective in eliminating any larger-scale spatio-temporal structures. Moreover, the associated species number anti-correlations prevent synchronous local coherent oscillatory modes.[35] As stochastic noise generates a marked attenuation of the population oscillations, whose frequencies are slightly shifted downwards relative to the mean-prediction ω_0, each species reaches its stationary density comparatively fast.[94] A perturbative analysis based on the corre-

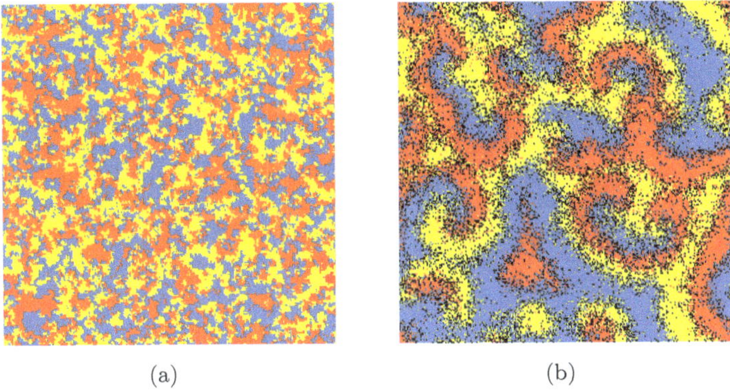

Figure 16.   Monte Carlo simulation snapshots of stochastic cyclic competition models on 256×256 square lattices with periodic boundary conditions, site occupation number exclusion, and random initial particle placement: (a) Cyclic Lotka–Volterra (rock-paper-scissors) model with conserved total particle number N; (b) May–Leonard model with separate predation and reproduction processes. Sites occupied by A, B, and C particles are respectively colored in red, yellow, and blue; black: empty spaces. [Figures reproduced with permission from (a) Ref. [94], copyright (2010) by The American Physical Society; (b) Ref. [99], copyright (2011) by EDP Sciences.]

sponding Doi–Peliti action qualitatively confirms these numerical findings. The fluctuation-induced damping coefficient γ is positive, at least to first order in λ/D, which indicates that the spatially uniform species distribution remains stable.[35]

Intriguingly though, in the extreme asymmetric predation rate limit $\lambda_A \gg \lambda_B, \lambda_C$, the rock-paper-scissors system effectively recovers the two-species Lotka–Volterra model with predators A, prey B, and abundant C population that essentially fills the system, $\langle n_C(\infty) \rangle \approx N$.[35,95] On the mean-field level, neglecting fluctuations in the C density yields the effective predator death rate $\mu = \langle n_C(\infty) \rangle \lambda_C$ and prey birth rate $\sigma = \langle n_C(\infty) \rangle \lambda_B$. A field-theoretical analysis confirms the mapping of the strongly asymmetric rock-paper-scissors Doi–Peliti action to its Lotka–Volterra counterpart (14), or equivalently to the effective Langevin equations (15) with noise correlations (16).[35] Indeed, lattice simulations of the rock-paper-scissors model with massive asymmetry in the predation rates display the characteristic Lotka–Volterra evasion-pursuit fronts and persistent population oscillations.[95]

6.2. *Spiral structure formation in the May–Leonard model*

The cyclic three-species May–Leonard model probably more realistically separates the predation $\sim \lambda$ and reproduction reactions $\sim \sigma$ to independent stochastic processes,[92] which lifts the constraint of total population conservation and induces a much richer phenomenology than in the cyclic Lotka–Volterra model.[23,24,26] It is defined by the competing reactions $A + B \xrightarrow{\lambda} A$ (and cyclic) and $A \underset{\kappa}{\overset{\sigma}{\rightleftharpoons}} A + A$, where the reverse reaction $\sim \kappa$ is set to limit the overall population, and where uniform rates across all three species have been assumed here. The corresponding mean-field rate equations read

$$\frac{\partial n_A(t)\rangle}{\partial t} = \langle n_A(t)\rangle \left[\sigma - \kappa\langle n_A(t)\rangle - \lambda\langle n_C(t)\rangle\right], \quad \text{(and cyclic)}, \qquad (23)$$

giving the symmetric three-species coexistence fixed point $\langle n_{A/B/C}(t)\rangle = \sigma/(\kappa + \lambda)$ with associated stability matrix

$$\mathbf{L} = -\frac{\sigma}{\kappa + \lambda} \begin{pmatrix} \kappa & 0 & \lambda \\ \lambda & \kappa & 0 \\ 0 & \lambda & \kappa \end{pmatrix}.$$

It has a negative real eigenvalue $-\sigma$ and the complex conjugate pair $-\sigma\left(2\kappa - \lambda \pm i\sqrt{3}\,\lambda\right)/2\left(\kappa + \lambda\right)$, which describe oscillations at frequency $\omega_0 = \sqrt{3}\,\sigma\,\lambda/2\left(\kappa + \lambda\right)$ whose amplitude decays for $2\kappa > \lambda$, but blows up for $2\kappa < \lambda$, indicating a Hopf bifurcation at $\kappa_c = \lambda/2$. For $\kappa > \kappa_c$, the mean-field coexistence fixed point is a stable focus, and approached in spiralling trajectories. Conversely in the unstable regime for $\kappa < \kappa_c$, the population trajectories in phase space assume the form of large-amplitude heteroclinic cycles, which in a stochastic system may reach the absorbing boundaries where either two species become extinct. The likelihood of fixation is exacerbated in strongly asymmetric stochastic May–Leonard systems.[96]

In a sufficiently large spatially extended May–Leonard model realization, the spontaneous emergence of spiral structures as depicted in Fig. 16(b) renders the system remarkably robust against single-species fixation,[97,98] even when model variations such as quenched disorder in the rates are introduced.[99,100] Yet as with increased particle mobility the characteristic spiral arm diameter approaches the linear exxtension L, the May–Leonard kinetics is driven unstable in finite systems and becomes prone to two-species extinction.[97] Near the Hopf bifurcation at κ_c, critical slowing down of the circularly polarized damped oscillating modes relative to the much faster (at rate σ) relaxing total particle density field affords a natural time scale separation in the system,[98] allowing one to integrate out the fast variable and map the rate equation model to the complex Ginzburg–Landau equation, one of the prime paradigms for spontaneous structure formation out of equilibrium.[101] However, for the fully stochastic May–Leonard model, the range of validity for this mapping becomes quite limited; through the non-vanishing noise cross-correlations, the fast mode is continuously excited, and one must resort to three stochastic fields to adequately capture the system's kinetics.[102] This can be achieved perturbatively via the associated Doi–Peliti path integral representation; yet the fluctuation corrections to lowest order do not alter the conclusions from the mean-field picture in a qualitative manner, but merely shift the Hopf bifurcation point and generate small oscillation parameter renormalizations.[35]

Generalization of the cyclic dominance motif to multiple competing species that may also form "alliances" depending on the structure of their interaction matrices has become a rich and intriguing research arena that can however not be adequately covered here.[26,103–114] It is worth mentioning though that an analysis of the associated adjacency matrix allows a remarkably complete classification of such many-species games with substructure within the mean-field theory realm, with decisive ramifications for the expected spontaneously forming patterns and ensuing coarsening kinetics in their spatially extended stochastic counterparts.[26,113]

6.3. *Inhomogeneous systems: diffusively coupled patches*

To conclude this discussion of stochastic spatial systems subject to binary predation processes, consider now spatially inhomogeneous settings where distinct "patches" controlled by different reaction rates are combined, and become effectively coupled through diffusive particle transport across their interfaces: As individuals cross the subsystem boundaries, they follow the stochastic processes and rules of their new environment. An example is illustrated in Fig. 17 for a series of successively finer-grained checkerboard squares of two-dimensional Lotka–Volterra predator prey patches alternatingly situated in the predator extinction and two-species coexistence regimes.[115] In the quasi-steady state, each subsystem has clearly evolved to its stationary prey fixation (uniformly green) or mixed-population configurations. Predators from the coexisting patches enter the fixation regimes, but the range of these immigrations remains constrained to thin boundary layers of the extent of a few lattice sites, set by the correlation length in the fixated region. As the number of subdivisions increases and the individual patch sizes reach the underlying lattice constant, the measured mean popu-

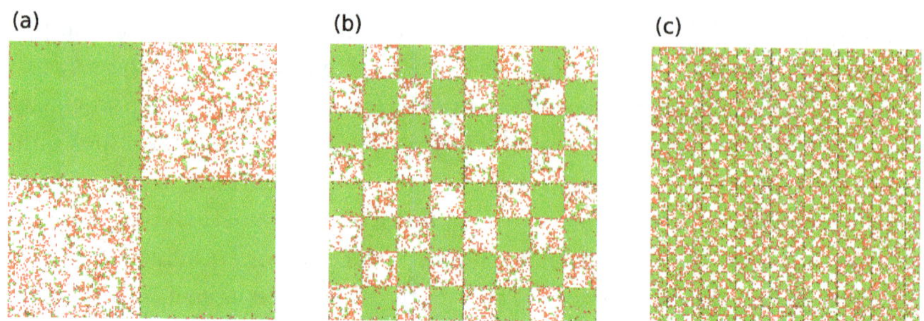

Figure 17. Simulation snapshots after 1000 MCS of the distribution of predators (red) and prey (green) in a stochastic Lotka–Volterra model on a spatially inhomogeneous 512×512 lattice (with periodic boundary conditions) with rates $\mu = 0.125$, $\sigma = 1$, and the predation rate set alternatingly in a checkerboard fashion between square patches with values $\lambda = 0.1$ (prey fixation) and $\lambda = 0.8$ (species coexistence). The system is split into successively smaller subdomains with linear sizes 256 (a), 64 (b), 16 (c). [Figures reproduced with permission from Ref. [115], copyright (2017) by Elsevier B.V.]

lation densities approach the corresponding results for a quenched random predation rate assignment drawn from a bimodal distribution.[115]

Fig. 18(a) schematically shows another realization of a diffusively coupled inhomogeneous setup, here for three-species cyclic rock-paper-scissors patches, with a thinner slab evolving under the cyclic Lotka–Volterra rules interacting with a much wider region governed by the May–Leonard reactions.[96,116] The small fluctuating clusters characteristic of the cyclic Lotka–Volterra model with conserved total particle number are visible on the left-hand side of Fig. 18(b). As the periodic population oscillations coherently impinge on the two interfaces, they generate planar invasion fronts moving into the larger May–Leonard region, which however decay on the scale of the characteristic correlation length ξ in that regime. In the interior of the May–Leonard patch, therefore its typical spiralling spatio-temporal patterns form, largely decoupled and independent from any influences from the rock-paper-scissors domain. In the vicinity of the interfaces, the population densities in the May–Leonard domain are reduced relative to the bulk, closer resembling the well-mixed mean-field values which do not account for the fitness enhancement due to the emergence of spiral structures.[116]

In both previous scenarios, stable subsystems, with stationary states that persist in the thermodynamic limit $L \to \infty$, interacted with each other through particle diffusion across the interfaces. The mutual influ-

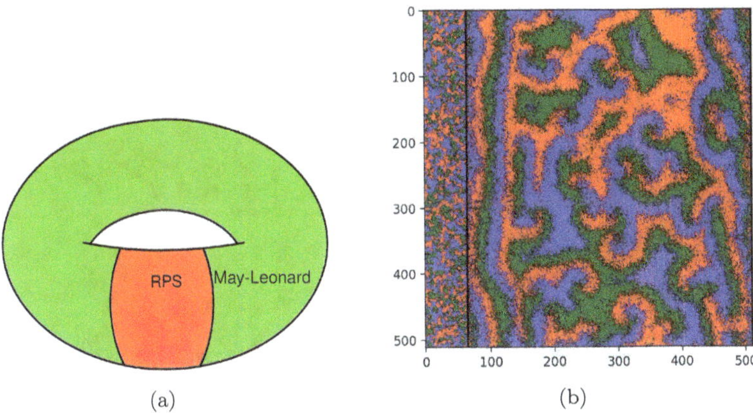

Figure 18. Spatially inhomogeneous rock-paper-scissors model realization with three cyclically competing species on a two-dimensional square lattice with periodic boundary conditions and site exclusion. (a) Illustration of the torus geometry and division into a smaller patch subject to cyclic Lotka–Volterra rules, and a larger region governed by the May–Leonard reactions. (b) Monte Carlo simulation snapshot in the (quasi-)stationary state for a 512×512 square lattice with the rock-paper-scissors patch of width 64 showing its characteristic fluctuating clusters that induce plane waves at both interfaces traveling into the seven times larger May–Leonard region ($\sigma = \lambda = 1$, $D = 0.5$). In the interior bulk of the May–Leonard subsystem one observes its typical spiral structures.[116]

ences are then limited to within a typical penetration depth set by the correlation length ξ, and thus affect only a finite boundary region. The situation changes drastically when one of the finite patches by itself would be unstable with respect to extinction or fixation. This finite-size instability with characteristic extinction or fixation time typically scaling exponentially with the particle number in the system $T \sim e^{cN}$ can then be suppressed, at least in an extended parameter regime, through a countermanding finite-size effect, namely the continuous supply of individuals from diverse species through the interface of area $\sim L^d$ that traverse the unstable patch in time $t_L \sim L/2v$ with invasion front speed v. If the particle influx is periodically driven by population oscillations at frequency ω in the stable adjacent regions, it may be roughly estimated as $v \sim \xi \omega / 2\pi$. A crude criterion for the induced stabilization of a vulnerable ecosystem would be that the particle import happens fast compared to the extinction time $t_L < T$, *i.e.*, the driving frequency should obey $\omega T > \pi L / \xi$.[117]

A striking example is depicted in Fig. 19 which shows two May–Leonard systems adjoined in the same geometric setting as in Fig. 18(a), where the

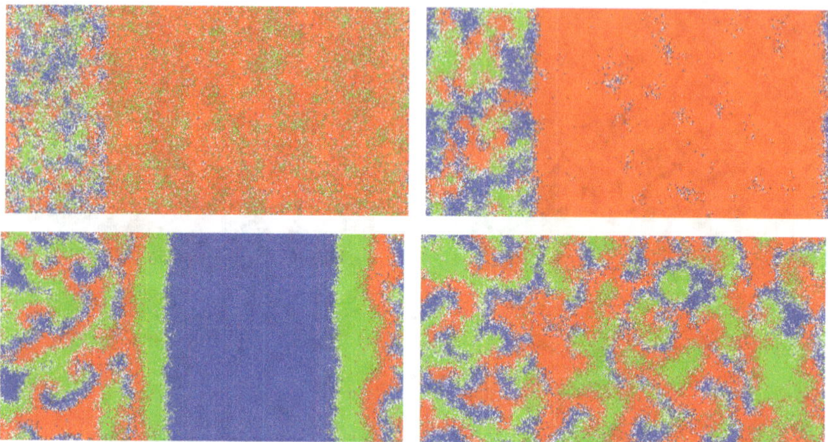

Figure 19. Monte Carlo simulation snapshots of a two-dimensional inhomogeneous May–Leonard system (periodic boundary conditions) where a (stable) 128 × 256 patch with symmetric rates $\sigma = \lambda = 0.2$ and diffusivity $D = 0.8$ is coupled via particle transport across the boundaries to a larger (unstable) 384 × 256 region with reduced reaction rate $\lambda' = 0.1$ for the process $A + B \to A$. At 100 MCS (top left), dominance of one (red) species has developed in the asymmetric patch; at 220 MCS (top right), spiraling clusters have formed in the smaller symmetric region; as time progesses, at 940 MCS (bottom left) planar waves emanate from the stable patch at both interfaces into the asymmetric area, by then fixated at the blue population; ultimately, in the quasi-stationary regime (bottom right), stable spiral structures are continuously seeded by the particle influx into the asymmetric area and persist throughout the simulation's duration. [Figures reproduced with permission from Ref. [96], copyright (2021) by EDP Sciences.]

smaller patch on the left-hand side is governed by symmetrically set rates, equal for all three competing species, while the three times larger subsystem on the right is regulated by an asymmetric choice of the predation rates, with $\lambda' = \lambda/2$ for the process $A + B \to A$ relative to the corresponding reactions for the other two species. The stable symmetric subsystem quickly develops the characteristic spiral structures in the course of the simulation. In contrast, the region with asymmetric predation rates displays pronounced heteroclinic cycles apparent in Fig. 19(b,c) and moreover is sufficiently small to be prone to two-species extinction. Indeed, in Fig. 19(c) its bulk region has effectively fixated at one (blue) population. However, planar invasion fronts are simultaneously excited at both interfaces to the stable May–Leonard subsystem, and quickly invade the entire region, eventually providing a sufficient supply of diverse individuals to stabilize the vulnerable ecology as seen in Fig. 19(d) which features the spiral patterns of the asymmetric system that would also emerge in much larger isolated settings. For this system, one may formulate the stabilization criterion that the induced spiral rotation frequency ω must exceed the characteristic inverse time scale for structure disruption owing to diffusive particle transport through the spiral arms, which should be proportional to the inverse square of the average correlation length for the three species, $\omega > D/\langle \xi^2 \rangle$.[96] Similar observations can be made in spatially inhomogeneous May–Leonard system combinations where the asymmetry is prescribed in the diffusivity rather than reactivity, as well as in diffusively coupled two-species Lotka–Volterra predator models where one patch is rendered finite-size unstable through increasing its predation rate or carrying capacity.[117]

7. Conclusions and outlook

The stochastic spatially extended Lotka–Volterra model for predator-prey competition and coexistence provides a paradigmatic example for the potentially severe quantitative as well as qualitative shortcomings of the (crudest) mean-field description in terms of deterministic coupled nonlinear rate equations in many nonequilibrium systems. Internal fluctuations as well as external variability in the model parameters are seen to have profound influence on the dynamics of the system. Perhaps most strikingly, intrinsic randomness in conjunction with spatio-temporal correlations induced by the binary reaction kinetics induce spontaneous pattern formation in dimensions $d \leq 4$. Hence, exploring fluctuation and correlation effects beyond the mean-field description constitutes an important task in physical, chemical, biological, ecological, and sociological models, as they may cause major changes to the dynamics and induce novel emergent phenomena. Even when the rate equations provide a qualitatively adequate picture, there usually will be quantitative parameter renormalizations in stochastic nonlinear dynamical systems.

A continuum field theory representation of the stochastic master equation represents a powerful analytical tool to properly incorporate fluctuation and correlation effects, but of course, explicit calculations can be performed only for comparatively simple models. Individual-based Monte Carlo simulations are an extremely useful tool to understand the temporal evolution, possible emergence of structures, and quantitative characterization of interacting stochastic many-particle systems, and may provide helpful guidance for further mathematical analysis. However, a full scan through the entire parameter space is rarely feasible, and aside from potential algorithmic artifacts, numerical data are naturally subject to statistical errors and finite-size limitations that cannot always be overcome within reasonable time frames and cost boundaries. The optimal strategy to enhance our understanding of complex systems in nature will therefore always be to utilize a combination of all state-of-the-art methods at our disposal.

Our exploration of spatial predator-prey models over the past two decades turned out vastly more intriguing than originally anticipated, revealing many unexpected phenomena and pertinent features connected to quite distinct physical or biological systems. The summary in this chapter inevitably focuses on our group's contributions and likely but inadvertently does injustice to many other researchers' important work, for which the author sincerely apologizes. This research topic covers a broad interdisciplinary area, which has been exciting to engage in, but which has also posed its specific challenges owing to the dispersal of the relevant literature and the distinct terminology and fundamental approaches in different scientific fields. There still remains much to be accomplished in refining our methods and conceptual grasp, and extending them to more complicated scenarios and especially observational and experimental realizations. I already look forward to seeing and learning about these novel future developments.

Acknowledgements

The author deeply appreciates the organizers' kind invitation to participate in the Ising Lectures series in May 2023. This book chapter is dedicated to the Ukrainian people and their valiant struggle for freedom in peace.

The author gratefully acknowledges his insightful and productive collaborators in the reported and related work: Lazarus Arnau, Matthew Asker, Timo Aspelmeier (deceased), Nicholas Butzin, John Cardy, Jacob Carroll, Sheng Chen, Lauren Childs, Udaya Sree Datla, Oliviér Deloubrière, Shengfeng Deng, Kenneth Distefano, Ulrich Dobramysl, Kim Forsten-Williams, Erwin Frey, Ivan Georgiev, Yadin Goldschmidt, Manoj Gopalakrishnan, Qian He, Bassel Heiba, Lluís Hernandez Navarro, Henk Hilhorst, Haye Hinrichsen, Martin Howard, Hans-Karl Janssen, Julianna Kelley, Wei Li, Weigang Liu, Jerôme Magnin, William Mather, Mauro Mobilia, Ruslan Mukhamadiarov, Riya Nandi, Michel Pleimling, Priyanka, Matthew Raum, Beth Reid, Alastair Rucklidge, Beate Schmittmann, Gunter Schütz, Franz Schwabl (deceased), Shannon Serrao, Sara Shabani, Mohamed Swailem, Reda Tiani, Steffen Trimper, Fréderic van Wijland, Ben Vollmayr-Lee, Mark Washenberger, Louie Hong Yao, Canon Zeidan, and Royce Zia.

This research was in part supported by the U.S. National Science Foundation, Division of Materials Research under Award No. DMR-0308548; the U.S. National Science Foundation, Division of Mathematical Sciences under current Award No. NSF DMS-2128587; the U.S. Department of Energy, Office of Basic Energy Sciences, Division of Materials Sciences and Engineering under Award No. DEFG02-09ER46613; and was sponsored by the U.S. Army Research Office and was accomplished under Grant No. W911NF17-1-0156. The views and conclusions contained in this document are those of the author and should not be interpreted as representing the official policies, either expressed or implied, of the Army Research Office or the U.S. Government.

References

1. H. Haken, *Synergetics – An Introduction*, Springer (Berlin, 1983).
2. J. D. Murray, *Mathematical Biology, Vols. I & II*, Springer (New York, 3rd ed. 2002 / 2003).
3. D. Neal, *Introduction to Population Biology*, Cambridge University Press (Cambridge, 2nd ed. 2018).
4. A. J. Lotka, Analytical note on certain rhythmic relations in organic systems, *Proceedings of the National Academy of Sciences* **6**, 410 (1920).
5. V. Volterra, Fluctuations in the abundance of a species considered mathematically, *Nature* **118**, 558 (1926).
6. R. M. May, *Stability and Complexity in Model Ecosystems*, Princeton University Press (Princeton, 1973).
7. J. Maynard-Smith, *Models in Ecology*, Cambridge University Press (Cambridge, 1974).
8. J. Maynard Smith, *Evolution and the Theory of Games*, Cambridge University Press (Cambridge, 1982).
9. J. Hofbauer and K. Sigmund, *Evolutionary Games and Population Dynamics*, Cambridge University Press (Cambridge, 1998).
10. U. C. Täuber, *Critical Dynamics — A Field Theory Approach to Equilibrium and Non-equilibrium Scaling Behavior*, Cambridge University Press (Cambridge, 2014).
11. U. C. Täuber, *Fluctuations and Correlations in Chemical Reaction Kinetics and Population Dynamics*, in Ref. [12], Chap. 1, 1.
12. K. Lindenberg, R. Metzler, and G. Oshanin (eds.), *Chemical Kinetics Beyond the Textbook*, World Scientific (Singapore, 2020).
13. U. Dobramysl, R. Mukhamadiarov, M. Swailem, and U. C. Täuber, Stochastic agent-based Monte Carlo simulations for reaction-diffusion models, population dynamics, and epidemic spreading, to be submitted to *American Journal of Physics* (2024).
14. V. Kuzovkov and E. Kotomin, Kinetics of bimolecular reactions in con-

densed media: Critical phenomena and microscopic self-organisation, *Rep. Prog. Phys.* **51**, 1479 (1988).

15. A. A. Ovchinnikov, S. F. Timashev, and A. A. Belyy, *Kinetics of Diffusion-Controlled Chemical Processes*, Nova Science (New York, 1989).

16. R. Durrett, Stochastic spatial models, *SIAM Review* **41**, 677 (1999).

17. H. Hinrichsen, Non-equilibrium critical phenomena and phase transitions into absorbing states, *Adv. Phys.* **49**, 815 (2000).

18. G. Ódor, Universality classes in nonequilibrium lattice systems, *Rev. Mod. Phys.* **76**, 663 (2004).

19. U. C. Täuber, M. Howard, and B. P. Vollmayr-Lee, Applications of field-theoretic renormalization group methods to reaction-diffusion problems, *J. Phys. A: Math. Gen.* **38**, R79 (2005); `arXiv:cond-mat/0501678`.

20. H. K. Janssen and U. C. Täuber, The field theory approach to percolation processes, *Ann. Phys.* (NY) **315**, 147 (2005); `arXiv:cond-mat/0409670`.

21. M. Mobilia, I. T. Georgiev, and U. C. Täuber, Phase transitions and spatio-temporal fluctuations in stochastic lattice Lotka–Volterra models, *J. Stat. Phys.* **128**, 447 (2007); `arXiv:q-bio.PE/0512039`.

22. M. Henkel, H. Hinrichsen, and S. Lübeck, *Non-equilibrium Phase Transitions, Vol. 1: Absorbing Phase Transitions*, Springer (Dordrecht, 2008).

23. E. Frey, Evolutionary game theory: Theoretical concepts and applications to microbial communities, *Physica A* **389**, 4265 (2010).

24. A. Szolnoki, M. Mobilia, L. L. Jiang, B. Szczesny, A. M. Rucklidge, and M. Perc, Cyclic dominance in evolutionary games: a review, *J. Roy. Soc. Interface* **11**, 20140735 (2014).

25. U. C. Täuber, Phase transitions and scaling in systems far from equilibrium, *Annu. Rev. Condens. Matter Phys.* **8**, 14 (2017).

26. U. Dobramysl, M. Mobilia, M. Pleimling, and U. C. Täuber, Stochastic population dynamics in spatially extended predator-prey systems, *J. Phys. A: Math. Theor.* **51**, 063001 (2018); `arXiv:1708.07055`.

27. N. G. van Kampen, *Stochastic Processes in Physics and Chemistry*, North Holland (Amsterdam, 1981).

28. C. M. van Vliet, *Equilibrium and Non-equilibrium Statistical Mechanics*, World Scientific (New Jersey, 2nd ed. 2010).

29. F. Schwabl, *Statistical Mechanics*, Springer (Berlin, 2nd ed. 2006).

30. M. Doi, Second quantization representation for classical many-particle systems, *J. Phys. A: Math. Gen.* **9**, 1465 (1976).

31. M. Doi, Stochastic theory of diffusion-controlled reactions, *J. Phys. A: Math. Gen.* **9**, 1479 (1976).

32. P. Grassberger and M. Scheunert, Fock-space methods for identical classical objects, *Fortschr. Phys.* **28**, 547 (1980).

33. L. Peliti, Path integral approach to birth-death processes on a lattice, *J. Phys.* (Paris) **46**, 1469 (1985).

34. U. C. Täuber, Population oscillations in spatial stochastic Lotka–Volterra models: A field-theoretic perturbational analysis, *J. Phys. A: Math. Theor.* **45**, 405002 (2012); `arXiv:1206.2303`.

35. L. H. Yao, M. Swailem, U. Dobramysl, and U. C. Täuber, Perturbative field-

theoretical analysis of three-species cyclic predator-prey models, *J. Phys. A: Math. Theor.* **56**, 225001 (2023); `arXiv:2303.08713`.

36. C. De Dominicis, Techniques de renormalisation de la théorie des champs et dynamique des phénomènes critiques, *J. Phys. (France) Colloq.* **C1**, C247 (1976).

37. H. K. Janssen, On a Lagrangean for classical field dynamics and renormalization group calculations of dynamical critical properties, *Z. Phys. B* **23**, 377 (1976).

38. R. Bausch, H. K. Janssen, and H. Wagner, Renormalized field theory of critical dynamics, *Z. Phys. B Cond. Matt.* **24**, 113 (1976).

39. M. Swailem and U. C. Täuber, Computing macroscopic reaction rates in reaction-diffusion systems using Monte Carlo simulations, *Phys. Rev. E* **110**, 014124 (2024); arXiv:2404.03089.

40. H. Matsuda, N. Ogita, A. Sasaki, and K. Satō, Statistical mechanics of population – the lattice Lotka–Volterra model, *Prog. Theor. Phys.* **88**, 1035 (1992).

41. J. E. Satulovsky and T. Tomé, Stochastic lattice gas model for a predator-prey system, *Phys. Rev. E* **49**, 5073 (1994).

42. N. Boccara, O. Roblin, and M. Roger, Automata network predator-prey model with pursuit and evasion, *Phys. Rev. E* **50**, 4531 (1994).

43. A. Provata, G. Nicolis, and F. Baras, Oscillatory dynamics in low-dimensional supports: a lattice Lotka–Volterra model, *J. Chem. Phys.* **110**, 8361 (1999).

44. A. F. Rozenfeld and E. V. Albano, Study of a lattice-gas model for a prey-predator system, *Physica A* **266**, 322 (1999).

45. A. Lipowski, Oscillatory behavior in a lattice prey-predator system, *Phys. Rev. E* **60**, 5179 (1999).

46. A. Lipowski and D. Lipowska, Nonequilibrium phase transition in a lattice prey-predator system, *Physica A* **276**, 456 (2000).

47. R. Monetti, A. Rozenfeld, and E. Albano, Study of interacting particle systems: The transition to the oscillatory behavior of a prey-predator model, *Physica A* **283**, 52 (2000).

48. M. Droz and A. Pękalski, Coexistence in a predator-prey system, *Phys. Rev. E* **63**, 051909 (2001).

49. T. Antal and M. Droz, Phase transitions and oscillations in a lattice prey-predator model, *Phys. Rev. E* **63**, 056119 (2001).

50. M. Kowalik, A. Lipowski, and A. L. Ferreira, Oscillations and dynamics in a two-dimensional prey-predator system, *Phys. Rev. E* **66**, 066107 (2002).

51. E. M. Rauch, H. Sayama, and Y. Bar-Yam, Dynamics and genealogy of strains in spatially extended host-pathogen models, *J. Theor. Biol.* **221**, 655 (2003).

52. M. Mobilia, I. T. Georgiev, and U. C. Täuber, Fluctuations and correlations in lattice models for predator-prey interaction, *Phys. Rev. E* **73**, 040903(R) (2006); `arXiv:q-bio.PE/0508043`.

53. M. J. Washenberger, M. Mobilia, and U. C. Täuber, Influence of local carrying capacity restrictions on stochastic predator-prey models, *J. Phys. Cond.*

Matt. **19**, 065139 (2007); arXiv:cond-mat/0606809.

54. U. Dobramysl and U. C. Täuber, Environmental versus demographic variability in stochastic predator–prey models, *J. Stat. Mech.* **2013**, P10001 (2013); arXiv:1307.4327.

55. M. Swailem and U. C. Täuber, Lotka–Volterra predator-prey model with periodically varying carrying capacity, *Phys. Rev. E* **107**, 064144 (2023); arXiv:2211.09276.

56. M. Parker and A. Kamenev, Mean extinction time in a predator–prey model, *J. Stat. Phys.* **141**, 201 (2009).

57. A. Dobrinevski and E. Frey, Extinction in neutrally stable stochastic Lotka–Volterra models, *Phys. Rev. E* **85**, 051903 (2012).

58. Movies of Monte Carlo simulation runs with various parameter settings can be viewed at: http://www.phys.vt.edu/~tauber/PredatorPrey/movies/.

59. T. Butler and N. Goldenfeld, Robust ecological pattern formation induced by demographic noise, *Phys. Rev. E* **80**, 030902(R) (2009).

60. T. Butler and N. Goldenfeld, Fluctuation-driven Turing patterns, *Phys. Rev. E* **84**, 011112 (2011).

61. U. Dobramysl and U. C. Täuber, Spatial variability enhances species fitness in stochastic predator-prey interactions, *Phys. Rev. Lett.* **101**, 258102 (2008); arXiv:0804.4127.

62. D. Toussaint and F. Wilczek, Particle-antiparticle annihilation in diffusive motion, *J. Chem. Phys.* **78**, 2642 (1983).

63. B. P. Lee and J. Cardy, Renormalization group study of the $A + B \to \emptyset$ diffusion-limited reaction, *J. Stat. Phys.* **80**, 971 (1995).

64. H. J. Hilhorst, O. Deloubrière, M. J. Washenberger, and U. C. Täuber, Segregation in diffusion-limited multispecies pair annihilation, *J. Phys. A: Math. Gen.* **37**, 7063 (2004).

65. H. J. Hilhorst, M. J. Washenberger, and U. C. Täuber, Symmetry and species segregation in diffusion-limited pair annihilation, *J. Stat. Mech.* **2004**, P10002 (2004).

66. A. J. McKane and T. J. Newman, Predator-prey cycles from resonant amplification of demographic stochasticity, *Phys. Rev. Lett.* **94**, 218102 (2005).

67. H. K. Janssen, On the nonequilibrium phase transition in reaction-diffusion systems with an absorbing stationary state, *Z. Phys. B Cond. Matt.* **42**, 15 (1981).

68. P. Grassberger, On phase transitions in Schlögl's second model, *Z. Phys. B Cond. Matt.* **47**, 365 (1982).

69. H. K. Janssen, Spontaneous symmetry breaking in directed percolation with many colors: Differentiation of species in the Gribov process, *Phys. Rev. Lett.* **78**, 2890 (1997).

70. H. K. Janssen, Directed percolation with colors and flavors, *J. Stat. Phys.* **103**, 801 (2001).

71. J. L. Cardy and R. L. Sugar, Directed percolation and Reggeon field theory, *J. Phys. A: Math. Gen.* **13**, L423 (1980).

72. S. Chen and U. C. Täuber, Non-equilibrium relaxation in a stochastic lattice Lotka–Volterra model, *Phys. Biol.* **13**, 025005 (2016); arXiv:1511.05114.

73. H. K. Janssen, B. Schaub, and B. Schmittmann, New universal short-time scaling behaviour of critical relaxation processes, *Z. Phys. B Cond. Matt.* **73**, 539 (1989).

74. B. Zheng, Generalized dynamic scaling for critical magnetic systems, *Int. J. of Mod. Phys. B* **12**, 1419 (1998).

75. P. Calabrese and A. Gambassi, Ageing properties of critical systems, *J. Phys. A: Math. Gen.* **38**, R133 (2005).

76. M. Henkel and M. Pleimling, *Non-equilibrium Phase Transitions, Vol. 2: Ageing and Dynamical Scaling Far from Equilibrium*, Springer (Dordrecht 2010).

77. H. Y. Shih, T. L. Hsieh, and N. Goldenfeld, Ecological collapse and the emergence of travelling waves at the onset of shear turbulence, *Nature Physics* **12**, 245 (2016).

78. K. A. Takeuchi, M. Kuroda, H. Chaté, and M. Sano, Directed percolation criticality in turbulent liquid crystals, *Phys. Rev. Lett.* **99**, 234503 (2007).

79. K. A. Takeuchi, M. Kuroda, H. Chaté, and M. Sano, Experimental realization of directed percolation criticality in turbulent liquid crystals, *Phys. Rev. E* **80**, 051116 (2009).

80. U. S. Datla, W. H. Mather, S. Chen, I. W. Shoultz, U. C. Täuber, C. N. Jones, and N. C. Butzin, The spatiotemporal system dynamics of acquired resistance in an engineered microecology, *Scientific Reports* **7**, 16071 (2017).

81. J. Kelley and U. C. Täuber, unpublished (2024).

82. S. Chen, U. Dobramysl, and U. C. Täuber, Evolutionary dynamics and competition stabilize three-species predator-prey communities, *Ecological Complexity* **36**, 57 (2018); arXiv:1711.05208.

83. U. Dobramysl and U. C. Täuber, Environmental versus demographic variability in two-species predator-prey models, *Phys. Rev. Lett.* **110**, 048105 (2013); arXiv:1206.0973.

84. W. O. Kermack, A. G. McKendrick, and G. T. Walker, A contribution to the mathematical theory of epidemics, *Proc. R. Soc. A* **115**, 700 (1927).

85. E. Frey, U. C. Täuber, and F. Schwabl, *Phys. Rev. E* **49**, 5058 (1994); arXiv:cond-mat/9404004.

86. R. I. Mukhamadiarov, S. Deng, S. R. Serrao, Priyanka, R. Nandi, L. H. Yao, and U. C. Täuber, Social distancing and epidemic resurgence in agent-based susceptible-infectious-recovered models, *Scientific Reports* **11**, 130 (2021); arXiv:2006.02552.

87. R. I. Mukhamadiarov, S. Deng, S. R. Serrao, Priyanka, L. M. Childs, and U. C. Täuber, Requirements for the containment of COVID-19 disease outbreaks through periodic testing, isolation, and quarantine, *J. Phys. A: Math. Theor.* **55**, 034001 (2022); medRxiv:2020.10.21.20217331.

88. P. Grassberger, On the critical behavior of the general epidemic process and dynamical percolation, *Math. Biosci.* **63**, 157 (1983).

89. H. K. Janssen, Renormalized field theory of dynamical percolation, *Z. Phys. B Cond. Matt.* **58**, 311 (1985).

90. J. L. Cardy and P. Grassberger, Epidemic models and percolation, *J. Phys. A: Math. Gen.* **18**, L267 (1985).

91. R. I. Mukhamadiarov and U. C. Täuber, Effects of lattice dilution on the nonequilibrium phase transition in the stochastic susceptible-infectious-recovered model, *Phys. Rev. E* **106**, 034132 (2022) arXiv:2206.03906.

92. R. M. May and W. J. Leonard, Nonlinear aspects of competition between three species, *SIAM J. Appl. Math.* **29**, 243 (1975).

93. M. Peltomäki and M. Alava, Three- and four-state rock-paper-scissors games with diffusion, *Phys. Rev. E* **78**, 031906 (2008).

94. Q. He, M. Mobilia, and U. C. Täuber, Spatial rock-paper-scissors models with inhomogeneous reaction rates, *Phys. Rev. E* **82**, 051909 (2010); arXiv:1004.5275.

95. Q. He, R. K. P. Zia, and U. C. Täuber, On the relationship between cyclic and hierarchical three-species predator-prey systems and the two-species Lotka–Volterra model, *Eur. Phys. J. B* **85**, 141 (2012); arXiv:1111.1674.

96. S. R. Serrao and U. C. Täuber, Stabilizing spiral structures and population diversity in the asymmetric May–Leonard model through immigration, *Eur. Phys. J. B* **94**, 175 (2021); arXiv:2105.08126.

97. T. Reichenbach, M. Mobilia, and E. Frey, Mobility promotes and jeopardizes biodiversity in rock-paper-scissors games, *Nature* **448**, 1046 (2007).

98. T. Reichenbach, M. Mobilia, and E. Frey, Self-organization of mobile populations in cyclic competition, *J. Theor. Biol.* **254**, 368 (2008).

99. Q. He, M. Mobilia, and U. C. Täuber, Coexistence in the two-dimensional May–Leonard model with random rates, *Eur. Phys. J. B* **82**, 97 (2011); arXiv:1101.4963.

100. D. Lamouroux, S. Eule, T. Geisel, and J. Nagler, Discriminating the effects of spatial extent and population size in cyclic competition among species, *Phys. Rev. E* **86**, 021911 (2012).

101. I. S. Aranson and L. Kramer, The world of the complex Ginzburg–Landau equation, *Rev. Mod. Phys.* **74**, 99 (2002).

102. S. R. Serrao and U. C. Täuber, A stochastic analysis of the spatially extended May–Leonard model, *J. Phys. A: Math. Theor.* **50**, 404005 (2017); arXiv:1706.00309.

103. G. Szabó and T. Czáran, Defensive alliances in spatial models of cyclical population interactions, *Phys. Rev. E* **64**, 042902 (2001).

104. G. Szabó, Competing associations in six-species predator–prey models, *J. Phys. A: Math. Gen.* **38**, 6689 (2005).

105. M. Perc, A. Szolnoki, and G. Szabó, Cyclical interactions with alliance-specificheterogeneous invasion rates, *Phys. Rev. E* **75**, 052102 (2007).

106. G. Szabó, A. Szolnoki, and I. Borsos, Self-organizing patterns maintained by competing associations in a six-species predator-prey model, *Phys. Rev. E* **77**, 041919 (2008).

107. P. P. Avelino, D. Bazeia, L. Losano, J. Menezes, and B. F. de Oliveira, Junctions and spiral patterns in generalized rock-paper-scissors models, *Phys. Rev. E* **86**, 036112 (2012).

108. A. Roman, D. Dasgupta, and M. Pleimling, Interplay between partnership formation and competition in generalized May–Leonard games, *Phys. Rev. E* **87**, 032148 (2013).

109. J. Vukov, A. Szolnoki, and D. Szabó, Diverging fluctuations in a spatial five-species cyclic dominance game, *Phys. Rev. E* **88**, 022123 (2013).

110. A. Dobrinevski, M. Alava, T. Reichenbach, and E. Frey, Mobility-dependent selection of competing strategy associations, *Phys. Rev. E* **89**, 012721 (2014).

111. P. P. Avelino, D. Bazeia, L. Losano, J. Menezes, and B. F. de Oliveira, Interfaces with internal structures in generalized rock-paper-scissors models, *Phys. Rev. E* **89**, 042710 (2014).

112. S. Mowlaei, A. Roman, and M. Pleimling, Spirals and coarsening patterns in the competition of many species: a complex Ginzburg–Landau approach, *J. Phys. A: Math. Theor.* **47**, 165001 (2014).

113. A. Roman, D. Dasgupta, and M. Pleimling, A theoretical approach to understand spatial organization in complex ecologies, *J. Theor. Biol.* **403**, 10 (2016).

114. D. Labavić and H. Meyer-Ortmanns, Rock-paper-scissors played within competing domains in predator-prey games, *J. Stat. Mech.* **2016**, 113402 (2016).

115. B. Heiba, S. Chen, and U. C. Täuber, Boundary effects on population dynamics in stochastic lattice Lotka–Volterra models, *Physica A* **491**, 582 (2018); `arXiv:1706.02567`.

116. M. L. Arnau, S. R. Serrao, and U. C. Täuber, unpublished (2020).

117. K. Distefano, C. Zeidan, M. Swailem, and U. C. Täuber, in preparation (2024).

Chapter 3

Statistical Mechanics and Artificial Neural Networks: Principles, Models, and Applications

Lucas Böttcher*

Dept. of Computational Science and Philosophy, Frankfurt School of Finance and Management, Adickesallee 32–34, 60322 Frankfurt am Main, Germany
Dept. of Medicine, University of Florida, Gainesville, FL, 32610, United States of America

Gregory Wheeler[†]

Dept. of Computational Science and Philosophy, Frankfurt School of Finance and Management, Adickesallee 32–34, 60322 Frankfurt am Main, Germany

The field of neuroscience and the development of artificial neural networks (ANNs) have mutually influenced each other, drawing from and contributing to many concepts initially developed in statistical mechanics. Notably, Hopfield networks and Boltzmann machines are versions of the Ising model, a model extensively studied in statistical mechanics for over a century. In the first part of this chapter, we provide an overview of the principles, models, and applications of ANNs, highlighting their connections to statistical mechanics and statistical learning theory.

Artificial neural networks can be seen as high-dimensional mathematical functions, and understanding the geometric properties of their loss landscapes (*i.e.*, the high-dimensional space on which one wishes to find extrema or saddles) can provide valuable insights into their optimization behavior, generalization abilities, and overall performance. Visualizing these functions can help us design better optimization methods and improve their generalization abilities. Thus, the second part of this chapter focuses on quantifying geometric properties and visualizing loss functions associated with deep ANNs.

*l.boettcher@fs.de, ORCID: `https://orcid.org/0000-0003-1700-1897`
[†]g.wheeler@fs.de, ORCID: `https://orcid.org/0000-0001-6868-3213`

Contents

1. Introduction 118
    1.1. The McCulloch–Pitts calculus and Hebbian learning 118
    1.2. The Perceptron 119
    1.3. Objective Bayes, entropy, and information theory 121
    1.4. Ising model 122
2. Hopfield network 123
3. Boltzmann machine learning 126
    3.1. Boltzmann machines 127
    3.2. Restricted Boltzmann machines 129
    3.3. Derivation of the learning algorithm 131
4. Loss landscapes of artificial neural networks 134
    4.1. Principal curvature in random projections 134
    4.2. Extracting curvature information 139
    4.3. Hessian directions 145
5. Conclusions 151
References 152

1. Introduction

There has been a long tradition in studying learning systems within statistical mechanics.[1–3] Some versions of artificial neural networks (ANNs) were, in fact, inspired by the Ising model that has been proposed in 1920 by Wilhelm Lenz as a model for ferromagnetism and solved in one dimension by his doctoral student, Ernst Ising, in 1924.[4–6] In the language of machine learning, the Ising model can be considered as a non-learning recurrent neural network (RNN).

Contributions to ANN research extend beyond statistical mechanics, however, requiring a broad range of scientific disciplines. Historically, these efforts have been marked by eras of interdisciplinary collaboration referred to as cybernetics,[7,8] connectionism,[9,10] artificial intelligence,[11,12] and machine learning.[13,14] To emphasize the importance of multidisciplinary contributions in ANN research, we will often highlight the research disciplines of individuals who have played pivotal roles in the advancement of ANNs.

1.1. *The McCulloch–Pitts calculus and Hebbian learning*

An influential paper by the neuroscientist Warren S. McCulloch and logician Walter Pitts published in 1943 is often regarded as inaugurating neural network research,[15] but in fact there was an active research community at that time working on the mathematics of neural activity.[16] McCulloch and Pitts's pivotal contribution was to join propositional logic and Alan

Turing's mathematical notion of computation[17] to propose a calculus for creating compositional representations of neural behavior. Their work has later on been picked up by the mathematician Stephen C. Kleene,[18] who made some of their work more accessible.[a]

Contemporaneously, the psychologist Donald O. Hebb formulated in his 1949 book "The Organization of Behavior"[19] a neurophysiological postulate, which is known today as "Hebb's rule" or "Hebbian learning":

> Let us assume then that the persistence or repetition of a reverberatory activity (or "trace") tends to induce lasting cellular changes that add to its stability. ... When an axon of cell A is near enough to excite a cell B and repeatedly or persistently takes part in firing it, some growth process or metabolic change takes place in one or both cells such that A's efficiency, as one of the cells firing B, is increased.
>
> Donald O. Hebb in "The Organization of Behavior" (1949)

In the modern literature, Hebbian learning is often summarized by the slogan "neurons wire together if they fire together".[20]

1.2. *The Perceptron*

In 1958, the psychologist Frank Rosenblatt developed the perceptron,[21,22] which is essentially a binary classifier that is based on a single artificial neuron and a Hebbian-type learning rule.[b] In parallel to Rosenblatt's work, the electrical engineers Bernard Widrow and Ted Hoff developed a similar ANN named ADALINE (Adaptive Linear Neuron).[23]

We show a schematic of a perceptron in Fig. 1. It receives inputs $x_1, x_2, \ldots, x_n$ that are weighted with $w_1, w_2, \ldots, w_n$. Within the summer (indicated by a Σ symbol), the artificial neuron computes $\sum_i w_i x_i + b$, where b is a bias term. This quantity is then used as input of a step activation function (indicated by a step symbol). The output of this function is y.

---

[a]Kleene comments on the work by McCulloch and Pitts in his 1956 article:[18] "The present memorandum is partly an exposition of the McCulloch–Pitts results; but we found the part of their paper which treats of arbitrary nets obscure; so we have proceeded independently here."

[b]The artificial neuron of this type is sometimes referred to as "McCulloch–Pitts neuron", owing to its connection to the foundational work of McCulloch and Pitts.[15] Some authors distinguish McCulloch–Pitts neurons from perceptrons by noting that the former handle binary signals, while the latter can process real-valued inputs.

               L. Böttcher and G. Wheeler

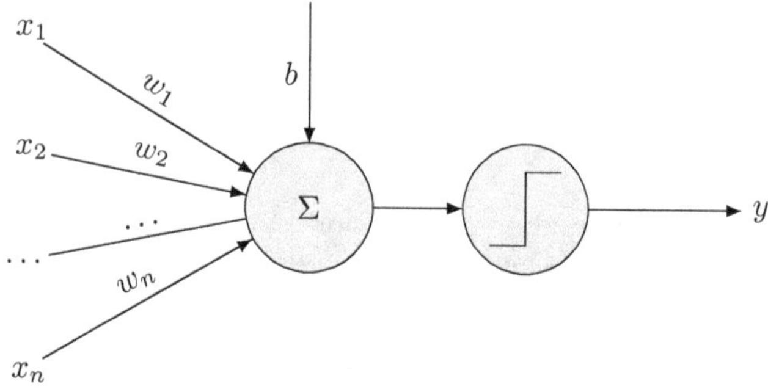

Figure 1. A schematic of a perceptron with output y, inputs $x_1, x_2, \ldots, x_n$, corresponding weights $w_1, w_2, \ldots, w_n$, and a bias term b. Traditionally, the inputs and outputs of perceptrons were considered to be binary (*e.g.*, $\{-1, 1\}$ or $\{0, 1\}$), while weights and biases are generally represented as real numbers. The symbol Σ signifies the summation process wherein the artificial neuron computes $\sum_i w_i x_i + b$. This computed value then becomes the input for a step activation function, denoted by a step symbol. We denote the final output as y. While the original perceptron employed a step activation function, contemporary applications frequently utilize a variety of functions including sigmoid, tanh, and rectified linear unit (ReLU).

Such simple artificial neurons can be used to implement logical functions including AND, OR, and NOT. For example, consider a perceptron with Heaviside activation, two binary inputs $x_1, x_2 \in \{0, 1\}$, weights $w_1, w_2 = 1$ and bias $b = -1.5$. We can readily verify that $y = x_1 \wedge x_2$. An example of a logical function that cannot be represented by a perceptron is XOR. This was shown by the computer scientists Marvin L. Minsky and Seymour A. Papert in their 1969 book "Perceptrons: An Introduction to Computational Geometry".[11] Based on their analysis of perceptrons, they considered the extension of research on this topic to be "sterile", implying that further exploration may not yield significant results.

Problems like the one associated with representing XOR using a single perceptron can be overcome by employing multilayer perceptrons (MLPs) or, in a broader sense, multilayer feedforward networks equipped with various types of activation functions.[24-26] One effective way to train such multilayered structures is reverse mode automatic differentiation (*i.e.*, backpropagation).[27-31]

Notice that perceptrons and other ANNs predominantly operate on real-valued data. Nevertheless, there is an expanding body of research focused on ANNs that are designed to handle complex-valued and, in some cases, quaternion-based information.[32-38]

1.3. *Objective Bayes, entropy, and information theory*

Over the same period of time, seminal work in probability and statistics appeared that would later prove crucial to the development of ANNs. In the early 20th century, a renewed interest in inverse "Bayesian" inference[39,40] met strong resistance from the emerging frequentist school of statistics led by Jerzy Neyman, Karl Pearson, and especially Ronald Fisher, who advocated his own *fiducial inference* as a better alternative to Bayesian inference.[41,42] In response to criticism that Bayesian methods were irredeemably subjective, Harold Jeffreys presented a methodology for "objectively" assigning prior probabilities based on symmetry principles and invariance properties.[43]

Edwin Jaynes, building on Jeffreys's ideas for objective Bayesian priors and Shannon's quantification of uncertainty,[44] proposed the Maximum Entropy (MaxEnt) principle as a rationality criteria for assigning prior probabilities that are at once consistent with known information while remaining maximally uncommitted about what remains uncertain.[45–47] Jaynes envisioned the MaxEnt principle as the basis for a "logic of science".[48] For instance, he proposed a reinterpretation of the principles of thermodynamics in information-theoretic terms where macro-states of a physical system are understood through the partial information they provide about the possible underlying microstates, showing how the original Boltzmann–Gibbs interpretation of entropy could be replaced by a more general single framework based on Shannon entropy. Objective Bayesianism does not remove subjective judgment,[49] MaxEnt is far from a universal principle for encoding prior information,[50] and the terms 'objective' and 'subjective' are outdated.[51,52]

For example, the MaxEnt principle can be used to select the least biased distribution from many common exponential family distributions by imposing constraints that match the expectation values of their sufficient statistics. That is, from constraints on the expected values of some function $h(x)$ on data parameterized by a vector θ, applying the MaxEnt principle to find the probability distribution that maximizes entropy under those constraints often takes the form of an exponential family distribution

$$p(x \mid \theta) = h(x) \exp\big(\theta^\top T(x) - A(\theta)\big) , \tag{1}$$

where $h(x)$ is the underlying base measure, θ is the natural (vector-valued) parameter of the distribution, $A(\theta)$ is the log normalizer, and $T(x)$ is the sufficient statistic determined by MaxEnt under the given constraints.[c]

---

[c]Here and in the following sections, we denote the transpose of a vector v as $v^\top$.

MaxEnt derived distributions depend on contingencies of the data and the availability of the right constraints to ensure the derivation goes through, however–contingencies that undermine viewing MaxEnt as a fundamental principle of rational inference.

Nevertheless, Jaynes's notion of treating probability as an extension of logic[48,48,53] and probabilistic inference as governed by coherence conditions on information states[54,55] has proved useful in optimization (*e.g.*, cross-entropy loss[56]), entropy-based regularization of ANNs,[57] and Bayesian neural networks.[58]

1.4. *Ising model*

Returning to the Ising model, a modified version of it has been endowed with a learning algorithm by the mathematical neuroscientist Shun'ichi Amari in 1972 and has been employed for learning patterns and sequences of patterns.[59] Amari's self-organizing ANN is an early model of associative memory that has later been popularized by the physicist John Hopfield.[60] Despite its limited representational capacity and use in practical applications, the study of Hopfield-type learning systems is still an active research area.[36,61–63]

The Hopfield network shares the Hamiltonian with the Ising model. Its learning algorithm is deterministic and aims at minimizing the "energy" of the system. One problem with the deterministic learning algorithm of Hopfield networks is that it cannot escape local minima. A solution to this problem is to employ a stochastic learning rule akin to Monte-Carlo methods that are used to generate Ising configurations.[64] In 1985, David H. Ackley, Geoffrey E. Hinton, and Terrence J. Sejnowski proposed such a learning algorithm for an Ising-type ANN which they called "Boltzmann machine" (BM).[65–68] In 1986, Paul Smolensky introduced the concept of "Harmonium",[69] which later evolved into restricted Boltzmann machines (RBMs). In contrast to Boltzmann machines, RBMs benefit from more efficient training algorithms that were developed in the early 2000s.[70,71] Since then, RBMs have been applied in various context such as to reduce dimensionality of datasets,[72] study phase transitions,[3,73–75] represent wave functions.[76,77]

Our historical overview of ANNs aims at emphasizing the strong interplay between statistical mechanics and related fields in the investigation of learning systems. Given the extensive span of over eight decades since the pioneering work of McCulloch and Pitts, summarizing the history of ANNs within a few pages cannot do justice to its depth and significance. We therefore refer the reader to the excellent review by Jürgen Schmidhuber (see Ref. [78]) for a more detailed overview of the history of deep learning in ANNs.

As computing power continues to increase, deep (*i.e.*, multilayer) ANNs have, over the past decade, become pervasive across various scientific domains.[79] Different established and newly developed ANN architectures have not only been employed to tackle scientific problems but have also found extensive applications in tasks such as image classification and natural language processing.[80–83]

Finally, we would also like to emphasize an area of research that has significantly advanced due to the availability of automatic differentiation[84] and the utilization of ANNs as universal function approximators.[85–87] This area resides at the intersection of non-linear dynamics,[88,89] control theory,[90–98] and dynamics-informed learning, where researchers often aim to integrate ANNs and their associated learning algorithms with domain-specific knowledge.[99–104] This integration introduces an inductive bias that facilitates effective learning.

The outline of this chapter is as follows. To better illustrate the connections between the Ising model and ANNs, we devote Secs. 2 and 3 to the Hopfield model and Boltzmann machines, respectively. In Sec. 4, we discuss how mathematical tools from high-dimensional probability and differential geometry can help us study the loss landscape of deep ANNs.[105] We conclude this chapter in Sec. 5.

2. Hopfield network

A Hopfield network is a complete (*i.e.*, fully connected) undirected graph in which each node is an artificial neuron of McCulloch–Pitts type, see Fig. 1. We show an example of a Hopfield network with $n = 5$ neurons in Fig. 2. We use $x_i \in \{-1, 1\}$ to denote the state of neuron i ($i \in \{1, \ldots, n\}$). Because the underlying graph is fully connected, neuron i receives $n - 1$ inputs x_j ($j \neq i$). The inputs x_j associated with neuron i are assigned weights $w_{ij} \in \mathbb{R}$. In a Hopfield network, weights are assumed to be symmetric (*i.e.*, $w_{ij} = w_{ji}$), and self-weights are considered to be absent (*i.e.*, $w_{ii} = 0$).

With these definitions in place, the activation of neuron i is given by

$$a_i = \sum_j w_{ij} x_j + b_i \,, \tag{2}$$

where b_i is the bias term of neuron i.

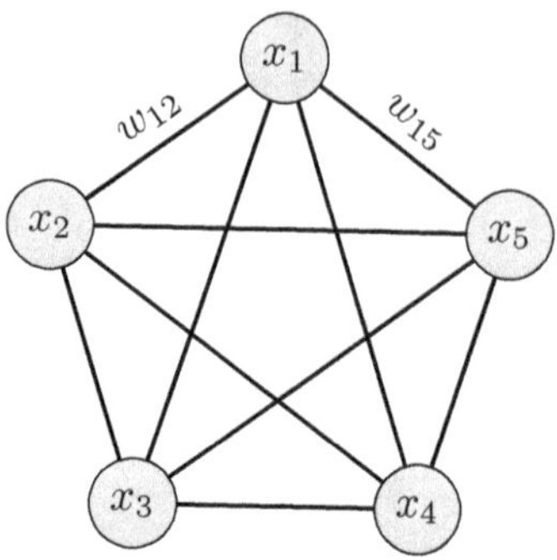

Figure 2.   An example of a Hopfield network with $n = 5$ neurons. Each neuron is connected to all other neurons through black edges, representing both inputs and outputs. (Adapted from Ref. [107].)

Regarding connections between the Ising model[106] and Hopfield networks, the states of artificial neurons align with the binary values employed in the Ising model to model the orientations of elementary magnets, often referred to as classical "spins". The weights in a Hopfield network represent the potentially diverse couplings between different elementary magnets. Lastly, the bias terms assume the function of external magnetic fields that act on these elementary magnets. Given these structural similarities between the Ising model and Hopfield networks, it is natural to assign them the Ising-type energy function

$$E = -\frac{1}{2} \sum_{i,j} w_{ij} x_i x_j - \sum_i b_i x_i \,. \tag{3}$$

Hopfield networks are not just a collection of interconnected artificial neurons, but they are dynamical systems whose states x_i evolve in discrete time according to

$$x_i = \begin{cases} 1\,, & \text{if } a_i \geq 0 \\ -1\,, & \text{otherwise}\,, \end{cases} \tag{4}$$

where a_i is the activation of neuron i, see Eq. (2).

For asynchronous updates in which the state of one neuron is updated at a time, the update rule (4) has an interesting property: Under this update rule, the energy E [see Eq. (3)] of a Hopfield network never increases.

The proof of this statement is straightforward. We are interested in the cases where an update of x_i causes this quantity to change its sign. Otherwise the energy will stay constant. Consider the two cases (i) $a_i =$

$\sum_j w_{ij} x_j + b_i < 0$ with $x_i = 1$, and (ii) $a_i \geq 0$ with $x_i = -1$. In the first case, the energy difference is

$$\Delta E_i = E(x_i = -1) - E(x_i = 1) = 2\left(b_i + \sum_j w_{ij} x_j\right) = 2a_i . \tag{5}$$

Notice that $\Delta E_i < 0$ because $a_i = \sum_j w_{ij} x_j + b_i < 0$.

In the second case, we have

$$\Delta E_i = E(x_i = 1) - E(x_i = -1) = -2\left(b_i + \sum_j w_{ij} x_j\right) = -2a_i . \tag{6}$$

The energy difference satisfies $\Delta E_i \leq 0$ because $a_i \geq 0$. In summary, we have shown that $\Delta E_i \leq 0$ for all changes of the state variable x_i according to the update rule (4).

Eqs. (5) and (6) reveal that the energy difference associated with a change of sign in x_i is $2a_i$ and $-2a_i$ for $a_i < 0$ and $a_i \geq 0$, respectively. Absorbing the bias term b_i in the weights w_{ij} by associating an extra active unit to every node in the network yields for the corresponding energy differences

$$\Delta E_i = \pm 2 \sum_j w_{ij} x_j . \tag{7}$$

One potential application of Hopfield networks is to store and retrieve information in local minima by adjusting their weights. For instance, consider the task of storing N binary (black and white) patterns in a Hopfield network. Let these N patterns be represented by the set $\{p_i^\nu = \pm 1 | 1 \leq i \leq n\}$ with $\nu \in \{1, \ldots, N\}$.

To learn the weights that represent our binary patterns, we can apply the Hebbian-type rule

$$w_{ij} = w_{ji} = \frac{1}{N} \sum_{\nu=1}^{N} p_i^\nu p_j^\nu . \tag{8}$$

In Hopfield networks with a large number of neurons, the capacity for storing and retrieving patterns is limited to approximately 14% of the total number of neurons.[108] However, it is possible to enhance this capacity through the use of modified versions of the learning rule (8).[109]

After applying the Hebbian learning rule to adjust the weights of a given Hopfield network, we may be interested in studying the evolution of different initial configurations $\{x_i\}$ according to update rule (4). A Hopfield network accurately represents a pattern $\{p_i^\nu = \pm 1 | 1 \leq i \leq n\}$ if x_i remains equal to

p_i^ν both before and after an update for all i (meaning $\{p_i^\nu\}$ is a fixed point of the system). Depending on the number of stored patterns and the chosen initial configuration, the Hopfield network may converge to a local minimum that does not align with the desired pattern. To mitigate this behavior, one can employ a stochastic update rule instead of the deterministic rule (4). This will be the topic that we are going to discuss in the following section.

3. Boltzmann machine learning

In Hopfield networks, if we start with an initial configuration that is close enough to a desired local energy minimum, we can recover the corresponding pattern $\{p_i^\nu\}$. However, for certain applications, relying solely on the deterministic update rule (4) may not be enough as it never accepts transitions that are associated with an increase in energy. In constraint satisfaction tasks, we often require learning algorithms to have the capability to occasionally accept transitions to configurations of higher energy and move the system under consideration away from local minima towards a global minimum. A method that is commonly used in this context is the $M(RT)^2$ algorithm introduced by Nicholas Metropolis, Arianna W. Rosenbluth, Marshall N. Rosenbluth, Augusta H. Teller, and Edward Teller in their seminal 1953 paper "Equation of State Calculations by Fast Computing Machines".[110,111] This algorithm became the basis of many optimization methods, such as simulated annealing.[112]

In the 1980s, the $M(RT)^2$ algorithm has been adopted to equip Hopfield-type systems with a stochastic update rule, in which neuron i is activated (set to 1) regardless of its current state, with probability

$$\sigma_i \equiv \sigma(\Delta E_i/T) = \sigma(2a_i/T) = \frac{1}{1 + \exp(-\Delta E_i/T)}. \tag{9}$$

Otherwise, it is set to -1. The corresponding ANNs were dubbed "'Boltzmann machines".[65–68]

In Eq. (9), the quantity $\Delta E_i = 2a_i$ is the energy difference between an inactive neuron i and an active one, see Eq. (5).[d] The function $\sigma(x) = 1/(1 + \exp(-x))$ represents the sigmoid function, and the parameter T serves as an equivalent to temperature.[e] In Fig. 3, we show $\sigma(\Delta E_i/T)$ as a function of ΔE_i for $T = 0.5, 1, 2$.

---

[d]We use the convention employed in Refs. [67,68]. The authors of Ref. [66], use the convention that $\Delta E_i = -2a_i$ and $\sigma_i \equiv \sigma(\Delta E_i/T) = 1/(1 + \exp(\Delta E_i/T))$.
[e]In the limit $T \to 0$, we recover the deterministic update rule (4).

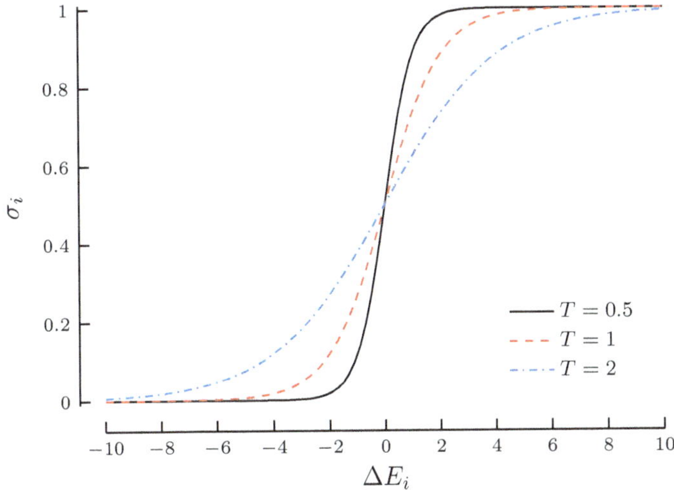

Figure 3.   The sigmoid function $\sigma_i \equiv \sigma(\Delta E_i/T)$ [see Eq. (9)] as function of ΔE_i for $T = 0.5, 1, 2$.

Examining Eqs. (3) and (9), we notice that we are simulating a system akin to the Ising model with Glauber (heat bath) dynamics.[64,113] Because Glauber dynamics satisfy the detailed balance condition, Boltzmann machines will eventually reach thermal equilibrium. The corresponding probabilities $p_{\mathrm{eq}}(X)$ and $p_{\mathrm{eq}}(Y)$ for the ANN to be in states X and Y, respectively, will satisfy[f]

$$\frac{p_{\mathrm{eq}}(Y)}{p_{\mathrm{eq}}(X)} = \exp\left(-\frac{E(Y) - E(X)}{T}\right). \tag{10}$$

In other words, the Boltzmann distribution provides the relative probability $p_{\mathrm{eq}}(Y)/p_{\mathrm{eq}}(X)$ associated with the states X and Y of a "thermalized" Boltzmann machine. Regardless of the initial configuration, at a given temperature T, the stochastic update rule in which neurons are activated with probability σ_i always leads to a thermal equilibrium configuration that is solely determined by its energy.

3.1. *Boltzmann machines*

Unlike Hopfield networks, Boltzmann machines have two types of nodes: visible units and hidden units. We denote the corresponding sets of visible

---

[f]We set the Boltzmann constant k_B to 1.

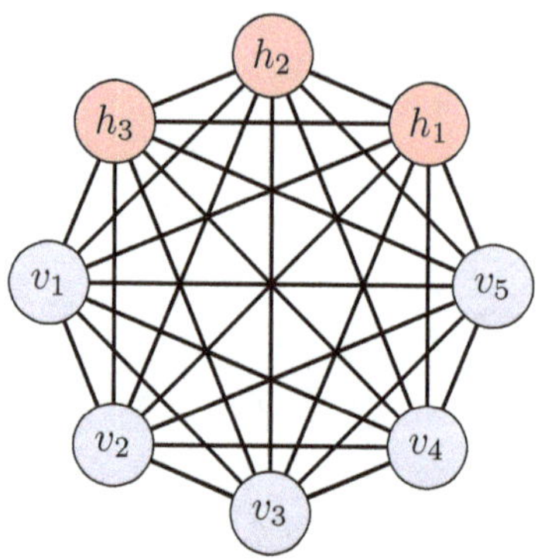

Figure 4. Boltzmann machines consist of visible units (blue) and hidden units (red). In the shown example, there are five visible units $\{v_i\}$ ($i \in \{1, \dots, 5\}$) and three hidden units $\{h_j\}$ ($j \in \{1, 2, 3\}$). Similar to Hopfield networks, the network architecture in a Boltzmann machine is complete.

and hidden units by V and H, respectively. Notice that the set H may be empty. In Fig. 4, we show an example of a Boltzmann machine with five visible units and three hidden units.

During the training of a Boltzmann machine, the visible units are "clamped" to the environment, which means they are set to binary vectors drawn from an empirical distribution. Hidden units may be used to account for constraints involving more than two visible units.

We denote the probability distribution of all configurations of visible units $\{\nu\}$ in a freely running network as $P'(\{\nu\})$. Here, "freely running" means that no external inputs are fixed (or "clamped") to the visible units. We derive the distribution $P'(\{\nu\})$ by summing (*i.e.*, marginalizing) over the corresponding joint probability distribution. That is,

$$P'(\{\nu\}) = \sum_{\{h\}} P'(\{\nu\}, \{h\}), \tag{11}$$

where the summation is performed over all possible configurations of hidden units $\{h\}$.

Our objective is now to devise a method such that $P'(\{\nu\})$ converges to the unknown environment (*i.e.*, data) distribution $P(\{\nu\})$. To do so, we quantify the disparity between $P'(\{\nu\})$ and $P(\{\nu\})$ using the Kullback–Leibler (KL) divergence (or relative entropy)

$$G(P, P') = \sum_{\{\nu\}} P(\{\nu\}) \ln \left[\frac{P(\{\nu\})}{P'(\{\nu\})} \right]. \tag{12}$$

To minimize the KL divergence $G(P, P')$, we perform a gradient descent according to

$$\frac{\partial G}{\partial w_{ij}} = -\frac{1}{T}\left(p_{ij} - p'_{ij}\right),\tag{13}$$

where p_{ij} represents the probability of both units i and j being active when the environment dictates the states of the visible units, and p'_{ij} is the corresponding probability in a freely running network without any connection to the environment.[66–68] We derive Eq. (13) in Sec. 3.3.

Both probabilities p_{ij} and p'_{ij} are measured once the Boltzmann machine has reached thermal equilibrium. Subsequently, the weights w_{ij} of the network are then updated according to

$$\Delta w_{ij} = \epsilon\left(p_{ij} - p'_{ij}\right),\tag{14}$$

where ϵ is the learning rate. To reach thermal equilibrium, the states of the visible and hidden units are updated using the update probability (9).

In summary, the steps relevant for training a Boltzmann machine are as follows.

> (1) Clamp the input data (environment distribution) to the visible units.
> (2) Update the state of all hidden units according to Eq. (9) until the system reaches thermal equilibrium and then compute p_{ij}.
> (3) Unclamp the input data from the visible units.
> (4) Update the state of all neurons according to Eq. (9) until the system reaches thermal equilibrium and then compute p'_{ij}.
> (5) Update all weights according to Eq. (14) and return to step 2 or stop if the weight updates are sufficiently small.

After training a Boltzmann machine, we can unclamp the visible units from the environment and generate samples to evaluate their quality. To do this, we use various initial configurations and activate neurons according to Eq. (9) until thermal equilibrium is reached. If the Boltzmann machine was trained successfully, the distribution of states for the unclamped visible units should align with the environment distribution.

3.2. *Restricted Boltzmann machines*

Boltzmann machines are not widely used in general learning tasks. Their impracticality arises from the significant computational burden associated

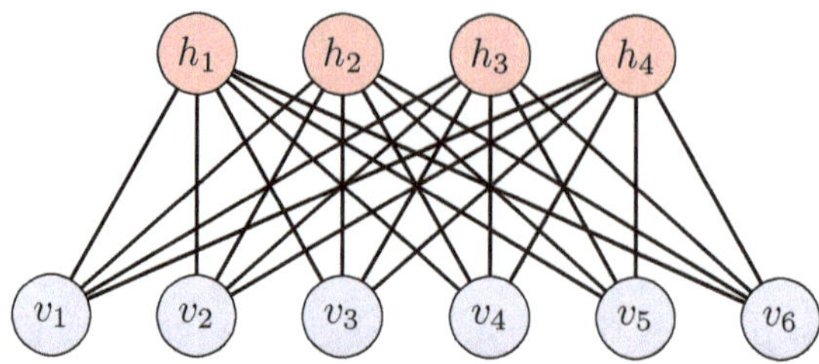

Figure 5.   Restricted Boltzmann machines consist of a visible layer (blue) and a hidden layer (red). In the shown example, the respective layers comprise six visible units $\{v_i\}$ ($i \in \{1,\ldots,6\}$) and four hidden units $\{h_j\}$ ($j \in \{1,\ldots,4\}$). The network structure of an RBM is bipartite and undirected.

with achieving thermal equilibrium, especially in instances involving large system sizes.

Restricted Boltzmann machines provide an ANN structure that can be trained more efficiently by omitting connections between the hidden and visible units, see Fig. 5. Because of these missing intra-layer connections, the network architecture of an RBM is bipartite.

In RBMs, updates for visible and hidden units are performed alternately. Because there are no connections within each layer, we can update all units within each layer in parallel. Specifically, visible unit v_i is activated with conditional probability

$$p(v_i = 1|\{h_j\}) = \sigma\left(\sum_j w_{ij} h_j + b_i\right), \tag{15}$$

where b_i is the bias of visible unit v_i and $\{h_j\}$ is a given configuration of hidden units. We then activate all hidden units based on their conditional probabilities

$$p(h_j = 1|\{v_i\}) = \sigma\left(\sum_i w_{ij} v_i + c_j\right), \tag{16}$$

where h_j represents hidden unit j, c_j is its associated bias, and $\{v_i\}$ denotes a specific configuration of visible units. This technique of sampling is referred to as "block Gibbs sampling".

Training an RBM shares similarities with training a BM. The key difference lies in the need to consider the bipartite network structure in the weight update Eq. (14). For an RBM, the weight updates are

$$\Delta w_{ij} = \epsilon(\langle \nu_i h_j \rangle_{\text{data}} - \langle \nu_i h_j \rangle_{\text{model}}). \tag{17}$$

Instead of sampling configurations to compute $\langle \nu_i h_j \rangle_{\text{data}}$ and $\langle \nu_i h_j \rangle_{\text{model}}$ at thermal equilibrium, we can instead employ a few relaxation steps. This approach is known as "contrastive divergence".[70,114] The corresponding weight updates are

$$\Delta w_{ij}^{\text{CD}} = \epsilon(\langle \nu_i h_j \rangle_{\text{data}} - \langle \nu_i h_j \rangle_{\text{model}}^k) \,. \tag{18}$$

The superscript k indicates the number of block Gibbs updates performed. For a more detailed discussion on the contrastive divergence training of RBMs, see Ref. [115].

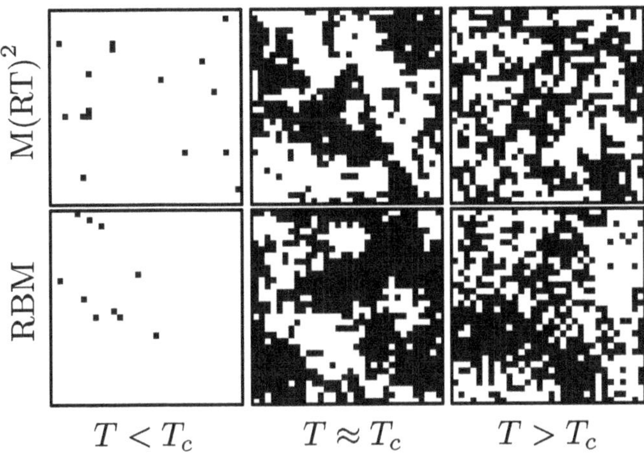

Figure 6. Snapshots of 32×32 Ising configurations are shown for $T \in \{1.5, 2.5, 4\}$. These configurations are derived from both M(RT)^2 and RBM samples. The quantity $T_c = 2/\ln(1 + \sqrt{2}) \approx 2.269$ is the critical temperature of the two-dimensional Ising model.

Restricted Boltzmann machines have been employed in diverse contexts, including dimensionality reduction of datasets,[72] the study of phase transitions,[3,73–75] and the representation of wave functions.[76,77] In Fig. 6, we show three snapshots of Ising configurations, generated using both M(RT)^2 sampling and RBMs, each comprising 32×32 spins. The RBM was trained using 20×10^4 realizations of Ising configurations sampled at various temperatures.[75]

3.3. *Derivation of the learning algorithm*

To derive Eq. (13), we follow the approach outlined in Ref. [68]. Notice that the environment distribution $P(\{\nu\})$ does not depend on w_{ij}. Hence, we have

$$\frac{\partial G}{\partial w_{ij}} = -\sum_{\{\nu\}} \frac{P(\{\nu\})}{P'(\{\nu\})} \frac{\partial P'(\{\nu\})}{\partial w_{ij}} . \tag{19}$$

Next, we wish to compute the gradient $\partial P'(\{\nu\})/\partial w_{ij}$. In a freely running BM, the equilibrium distribution of the visible units follows a Boltzmann distribution. That is,

$$P'(\{\nu\}) = \sum_{\{h\}} P'(\{\nu\}, \{h\}) = \frac{\sum_{\{h\}} e^{-E(\{\nu,h\})/T}}{\sum_{\{\nu,h\}} e^{-E(\{\nu,h\})/T}} . \tag{20}$$

Here, the quantity

$$E(\{\nu, h\}) = -\frac{1}{2} \sum_{i,j} w_{ij} x_i^{\{\nu,h\}} x_j^{\{\nu,h\}} \tag{21}$$

is the Ising-type energy function in which field (or bias) terms are absorbed in the weights w_{ij}, see Eqs. (3) and (7). For a BM in state $\{\nu, h\}$, we denote the state of neuron i by $x_i^{\{\nu,h\}}$. Using

$$\frac{\partial e^{-E(\{\nu,h\})/T}}{\partial w_{ij}} = \frac{1}{T} x_i^{\{\nu,h\}} x_j^{\{\nu,h\}} e^{-E(\{\nu,h\})/T} \tag{22}$$

yields

$$\begin{aligned}
\frac{\partial P'(\{\nu\})}{\partial w_{ij}} = & \frac{\frac{1}{T}\sum_{\{h\}} x_i^{\{\nu,h\}} x_j^{\{\nu,h\}} e^{-E(\{\nu,h\})/T}}{\sum_{\{\nu,h\}} e^{-E(\{\nu,h\})/T}} \\
& - \frac{\sum_{\{h\}} e^{-E(\{\nu,h\})/T} \frac{1}{T}\sum_{\{\nu,h\}} x_i^{\{\nu,h\}} x_j^{\{\nu,h\}} e^{-E(\{\nu,h\})/T}}{\left(\sum_{\{\nu,h\}} e^{-E(\{\nu,h\})/T}\right)^2} \\
= & \frac{1}{T} \left[\sum_{\{h\}} x_i^{\{\nu,h\}} x_j^{\{\nu,h\}} P'(\{\nu\}, \{h\}) \right. \\
& \left. - P'(\{\nu\}) \sum_{\{\nu,h\}} x_i^{\{\nu,h\}} x_j^{\{\nu,h\}} P'(\{\nu\}, \{h\}) \right] .
\end{aligned} \tag{23}$$

We will now substitute this result into Eq. (19) to obtain

$$\frac{\partial G}{\partial w_{ij}} = -\sum_{\{\nu\}} \frac{P(\{\nu\})}{P'(\{\nu\})} \frac{1}{T} \left[\sum_{\{h\}} x_i^{\{\nu,h\}} x_j^{\{\nu,h\}} P'(\{\nu\},\{h\}) \right.$$

$$\left. -P'(\{\nu\}) \sum_{\{\nu,h\}} x_i^{\{\nu,h\}} x_j^{\{\nu,h\}} P'(\{\nu\},\{h\}) \right]$$

$$= -\frac{1}{T} \left[\sum_{\{\nu,h\}} x_i^{\{\nu,h\}} x_j^{\{\nu,h\}} P(\{\nu\},\{h\}) \right.$$

$$\left. - \sum_{\{\nu,h\}} x_i^{\{\nu,h\}} x_j^{\{\nu,h\}} P'(\{\nu\},\{h\}) \right], \qquad (24)$$

where we used that $\sum_{\{\nu\}} P(\{\nu\}) = 1$ and $P(\{\nu\})/P'(\{\nu\}) P'(\{\nu\},\{h\}) = P(\{\nu\},\{h\})$. The latter equation follows from the definition of joint probability distributions

$$P(\{\nu\},\{h\}) = P(\{h\}|\{\nu\}) P(\{\nu\}) \qquad (25)$$

and

$$P'(\{\nu\},\{h\}) = P'(\{h\}|\{\nu\}) P'(\{\nu\}), \qquad (26)$$

with $P(\{h\}|\{\nu\}) = P'(\{h\}|\{\nu\})$. The conditional distributions $P(\{h\}|\{\nu\})$ and $P'(\{h\}|\{\nu\})$ are identical, as the probability of observing a certain hidden state given a visible one is independent of the origin of the visible state in an equilibrated system. In other words, for the conditional distributions in an equilibrated system, it is irrelevant whether the visible state is provided by the environment or generated by a freely running machine. Defining

$$p_{ij} := \sum_{\{\nu,h\}} x_i^{\{\nu,h\}} x_j^{\{\nu,h\}} P(\{\nu\},\{h\}) \qquad (27)$$

and

$$p'_{ij} := \sum_{\{\nu,h\}} x_i^{\{\nu,h\}} x_j^{\{\nu,h\}} P'(\{\nu\},\{h\}), \qquad (28)$$

we finally obtain Eq. (13).

4. Loss landscapes of artificial neural networks

Training an ANN involves minimizing a given loss function such as the KL divergence, see Eq. (12). The loss landscapes of ANNs are affected by several factors, including structural properties[86,87,116] and a range of implementation attributes.[117–119] Knowledge of their precise effects on learning performance, however, remains incomplete.

One path to a better understanding of the relationships between ANN structure, implementation attributes, and learning performance is through a more in-depth analysis of the geometric properties of loss landscapes. For instance, Keskar *et al.*[120] analyze the local curvature around candidate minimizers via the spectrum of the underlying Hessian to characterize the flatness and sharpness of loss minima, and Dinh *et al.*[121] demonstrate that reparameterizations can render flat minima sharp without affecting generalization properties. Even so, one challenge to the study of geometric properties of loss landscapes is high dimensionality. To meet that challenge, some propose to visualize the curvature around a given point by projecting curvature properties of high-dimensional loss functions to a lower-dimensional (and often random) projection of two or three dimensions.[122–126] Horoi *et al.*,[126] building on this approach, pursue improvements in learning by dynamically sampling points in projected low-loss regions surrounding local minima during training.

However, visualizing high-dimensional loss landscape curvature relies on accurate projections of curvature properties to lower dimensions. Unfortunately, random projections do not preserve curvature information, thus do not afford accurate low-dimensional representations of curvature information. This argument is given in Sec. 4.1 and illustrated with a simulation example in Sec. 4.2. Principal curvatures in a low-dimensional projection are nevertheless given by functions of weighted ensemble means of the Hessian elements in the original, high-dimensional space, a result also established in Sec. 4.1. Instead of using random projections to visualize loss functions, we propose to analyze projections along *dominant Hessian directions* associated with the largest-magnitude positive and negative principal curvatures.

4.1. *Principal curvature in random projections*

To describe the connection between the principal curvature of a loss function $L(\theta)$ with ANN parameters $\theta \in \mathbb{R}^N$ and that associated with a lower-dimensional, random projection, we provide in Sec. 4.1.1 an overview of concepts from differential geometry[127–129] that are useful to mathematically describe curvature in high-dimensional spaces. In Secs. 4.1.2 and 4.1.3, we show the relationship between principal curvature and random projections. Finally, in Sec. 4.1.4, we highlight a relationship between measures of curvature presented in Sec. 4.1.3 and Hessian trace estimates, which affords an alternative to Hutchinson's method for computing unbiased Hessian trace estimates.

4.1.1. *Differential and information geometry concepts*

In differential geometry, the principal curvatures are the eigenvalues of the shape operator (or Weingarten map[g]), a linear endomorphism defined on the tangent space T_p of L at a point p. For the original high-dimensional space, we have $(\theta, L(\theta)) \subseteq \mathbb{R}^{N+1}$ and there are N principal curvatures $\kappa_1^\theta \geq \kappa_2^\theta \geq \cdots \geq \kappa_N^\theta$. At a non-degenerate critical point θ^* where the gradient $\nabla_\theta L = (\partial L/\partial\theta_1, \ldots, \partial L/\partial\theta_N)^\top$ vanishes, the matrix of the shape parameter is given by the Hessian H_θ with elements $(H_\theta)_{ij} = \partial^2 L/(\partial\theta_i \partial\theta_j)$ $(i, j \in \{1, \ldots, N\})$.[h] Some works refer to the Hessian as the "curvature matrix"[131] or use it to characterize the curvature properties of $L(\theta)$.[123] In the vicinity of a non-degenerate critical point θ^*, the eigenvalues of the Hessian H_θ are the principal curvatures and describe the loss function in the eigenbasis of H_θ according to

$$L(\theta^* + \Delta\theta) = L(\theta^*) + \frac{1}{2}\sum_{i=1}^{N} \kappa_i^\theta \Delta\theta_i^2 \,. \tag{29}$$

The Morse lemma states that, if a critical point θ^* of $L(\theta)$ is non-degenerate, then there exists a chart $(\tilde\theta_1, \ldots, \tilde\theta_N)$ in a neighborhood of θ^* such that

$$L(\tilde\theta) = -\tilde\theta_1^2 - \cdots - \tilde\theta_i^2 + \tilde\theta_{i+1}^2 + \cdots + \tilde\theta_N^2 + L(\theta^*)\,, \tag{30}$$

where $\tilde\theta_i(\theta) = 0$ for $i \in \{1, \ldots, N\}$. The loss function $L(\tilde\theta)$ in Eq. (30) is decreasing along i directions and increasing along the remaining $i + 1$ to N directions. Further, the index i of a critical point θ^* is the number of negative eigenvalues of the Hessian H_θ at that point.

---

[g]Some authors distinguish between the shape operator and the Weingarten map depending on if the change of the underlying tangent vector is described in the original manifold or in Euclidean space (see, *e.g.*, chapter 3 in Ref. [130]).

[h]A critical point is degenerate if the Hessian H_θ at this point is singular (*i.e.*, $\det(H_\theta) = 0$). At degenerate critical points, one cannot use the eigenvalues of H_θ to determine if the critical point is a minimum (positive definite H_θ) or a maximum (negative definite H_θ). Geometrically, at a degenerate critical point, a quadratic approximation fails to capture the local behavior of the function that one wishes to study.

In the standard basis, the Hessian is

$$H_\theta := \nabla_\theta \nabla_\theta L(\theta) = \begin{pmatrix} \frac{\partial^2 L}{\partial \theta_1^2} & \cdots & \frac{\partial^2 L}{\partial \theta_1 \partial \theta_N} \\ \vdots & \ddots & \vdots \\ \frac{\partial^2 L}{\partial \theta_N \partial \theta_1} & \cdots & \frac{\partial^2 L}{\partial \theta_N^2} \end{pmatrix}. \tag{31}$$

4.1.2. *Random projections*

To graphically explore an N-dimensional loss function L around a critical point θ^*, one may wish to work in a lower-dimensional representation. For example, a two-dimensional projection of L around θ^* is provided by

$$L(\theta^* + \alpha\eta + \beta\delta), \tag{32}$$

where the parameters $\alpha, \beta \in \mathbb{R}$ scale the directions $\eta, \delta \in \mathbb{R}^N$. The corresponding graph representation is $(\alpha, \beta, L(\alpha, \beta)) \subseteq \mathbb{R}^3$.

In high-dimensional spaces, there exist vastly many more almost-orthogonal than orthogonal directions. In fact, if the dimension of our space is large enough, with high probability, random vectors will be sufficiently close to orthogonal.[132] Following this result, many related works[123,125,126] use random Gaussian directions with independent and identically distributed vector elements $\eta_i, \delta_i \sim \mathcal{N}(0, 1)$ $(i \in \{1, \ldots, N\})$.[i]

The scalar product of random Gaussian vectors η, δ is a sum of the difference between two chi-squared distributed random variables because

$$\sum_{i=1}^{N} \eta_i \delta_i = \sum_{i=1}^{N} \frac{1}{4}(\eta_i + \delta_i)^2 - \frac{1}{4}(\eta_i - \delta_i)^2 = \sum_{i=1}^{N} \frac{1}{2}X_i^2 - \frac{1}{2}Y_i^2, \tag{33}$$

where $X_i, Y_i \sim \mathcal{N}(0, 1)$.

Notice that η, δ are almost orthogonal, which can be proved using a concentration inequality for chi-squared distributed random variables.[j]

4.1.3. *Principal curvature*

With the form of random Gaussian projections in hand, we now analyze the principal curvatures in both the original and lower-dimensional spaces.

---

[i]Here, $\eta_i, \delta_i \sim \mathcal{N}(0, 1)$ means that the elements η_i, δ_i of the vectors η, δ are distributed according to a standard normal distribution $\mathcal{N}(0, 1)$.
[j]For further details, see Ref. [105].

The Hessian associated with the two-dimensional loss projection (32) is

$$
H_{\alpha,\beta} = \begin{pmatrix} \frac{\partial^2 L}{\partial \alpha^2} & \frac{\partial^2 L}{\partial \alpha \partial \beta} \\ \frac{\partial^2 L}{\partial \beta \partial \alpha} & \frac{\partial^2 L}{\partial \beta^2} \end{pmatrix} = \begin{pmatrix} \sum_{i,j} \eta_i \eta_j \frac{\partial^2 L}{\partial \theta_i \theta_j} & \sum_{i,j} \eta_i \delta_j \frac{\partial^2 L}{\partial \theta_i \theta_j} \\ \sum_{i,j} \eta_i \delta_j \frac{\partial^2 L}{\partial \theta_i \theta_j} & \sum_{i,j} \delta_i \delta_j \frac{\partial^2 L}{\partial \theta_i \theta_j} \end{pmatrix}
$$
$$
= \begin{pmatrix} (H_\theta)_{ij} \eta^i \eta^j & (H_\theta)_{ij} \eta^i \delta^j \\ (H_\theta)_{ij} \eta^i \delta^j & (H_\theta)_{ij} \delta^i \delta^j \end{pmatrix},
\tag{34}
$$

where we use Einstein notation in the last equality.

Because the elements of δ, η are distributed according to a standard normal distribution, the second derivatives of the loss function L in Eq. (34) have prefactors that are products of standard normal variables and, hence, can be expressed as sums of chi-squared distributed random variables as in Eq. (33). To summarize, elements of $H_{\alpha,\beta}$ are sums of second derivatives of L in the original space weighted with chi-squared distributed prefactors.

The principal curvatures $\kappa_\pm^{\alpha,\beta}$ (*i.e.*, the eigenvalues of $H_{\alpha,\beta}$) are

$$
\kappa_\pm^{\alpha,\beta} = \frac{1}{2} \left(A + C \pm \sqrt{4B^2 + (A - C)^2} \right),
\tag{35}
$$

where $A = (H_\theta)_{ij} \eta^i \eta^j$, $B = (H_\theta)_{ij} \eta^i \delta^j$, and $C = (H_\theta)_{ij} \delta^i \delta^j$. To the best of our knowledge, a closed, analytic expression for the distribution of the quantities A, B, C is not yet known.[133–136]

Returning to principal curvature, since $\sum_{i,j} a_{ij} \eta^i \eta^j = \sum_i a_{ii} \eta^i \eta^i + \sum_{i \neq j} a_{ij} \eta^i \eta^j$ ($a_{ij} \in \mathbb{R}$), we find that $\mathbb{E}[A] = \mathbb{E}[C] = (H_\theta)^i{}_i$ and $\mathbb{E}[B] = 0$ where $(H_\theta)^i{}_i \equiv \mathrm{tr}(H_\theta) = \sum_{i=1}^N \kappa_i^\theta$.[k] To show that the expected values of $a_{ij} \eta^i \eta^j$ ($i \neq j$) or $a_{ij} \eta^i \delta^j$ vanish, one can either invoke independence of η^i, η^j ($i \neq j$) and η^i, δ^j or transform both products into corresponding differences of two chi-squared random variables with the same mean, see Eq. (33).

Hence, the expected, dimension-reduced Hessian (34) is

$$
\mathbb{E}[H_{\alpha,\beta}] = \begin{pmatrix} (H_\theta)^i{}_i & 0 \\ 0 & (H_\theta)^i{}_i \end{pmatrix}.
\tag{36}
$$

The corresponding eigenvalue (or principal curvature) $\bar{\kappa}^{\alpha,\beta}$ is therefore given by the sum over all principal curvatures in the original space (*i.e.*, $\bar{\kappa}^{\alpha,\beta} = \sum_{i=1}^N \kappa_i^\theta$). Hence, the value of the principal curvature $\bar{\kappa}^{\alpha,\beta}$ in the expected dimension-reduced space will be either positive (if the positive

---

[k]The expected values of the quantities A, B, and C correspond to ensemble means (43) in the limit $S \to \infty$, where S is the number of independent realizations of the underlying random variable.

Table 1. Quantities to characterize curvature.

SYMBOL	DEFINITION
$H_\theta \in \mathbb{R}^{N \times N}$	Hessian in original loss space
$\kappa_i^\theta \in \mathbb{R}$	principal curvatures in original loss space with $i \in \{1, \dots, N\}$ (*i.e.*, the eigenvalues of H_θ)
$H_{\alpha,\beta} \in \mathbb{R}^{2 \times 2}$	Hessian in a two-dimensional projection of an N-dimensional loss function
$\kappa_\pm^{\alpha,\beta} \in \mathbb{R}$	principal curvatures in a two-dimensional loss projection (*i.e.*, the eigenvalues of $H_{\alpha,\beta}$)
$\bar{\kappa}^{\alpha,\beta} \in \mathbb{R}$	principal curvature in expected, two-dimensional loss projection (*i.e.*, the eigenvalues of $\mathbb{E}[H_{\alpha,\beta}]$)
$H \in \mathbb{R}$	mean curvature (*i.e.*, $\sum_{i=1}^N \kappa_i^\theta / N = \bar{\kappa}^{\alpha,\beta}/N$)

principal curvatures in the original space dominate), negative (if the negative principal curvatures in the original space dominate), or close to zero (if positive and negative principal curvatures in the original space cancel out each other). As a result, saddle points will not appear as such in the expected random projection.

In addition to the connection between $\bar{\kappa}^{\alpha,\beta}$ and the principal curvatures κ_i^θ, we now provide an overview of additional mathematical relations between different curvature measures that are useful to quantify curvature properties of high-dimensional loss functions and their two-dimensional random projections.

Invoking Eq. (35), we can relate $\bar{\kappa}^{\alpha,\beta}$ to $\mathrm{tr}(H_\theta)$ and $\kappa_\pm^{\alpha,\beta}$. Because $\kappa_+^{\alpha,\beta} + \kappa_-^{\alpha,\beta} = A + C$, we have

$$\mathrm{tr}(H_\theta) = \bar{\kappa}^{\alpha,\beta} = \sum_{i=1}^N \kappa_i^\theta = \frac{1}{2}\left(\mathbb{E}[\kappa_-^{\alpha,\beta}] + \mathbb{E}[\kappa_+^{\alpha,\beta}]\right). \tag{37}$$

The mean curvature H in the original space is related to $\bar{\kappa}^{\alpha,\beta}$ via

$$H = \frac{1}{N}\bar{\kappa}^{\alpha,\beta} = \frac{1}{N}\sum_{i=1}^N \kappa_i^\theta. \tag{38}$$

We summarize the definitions of the employed Hessians and curvature measures in Tab. 1.

The appeal of random projections is that pairwise distances between points in a high-dimensional space can be nearly preserved by a lower-dimensional linear embedding, affording a low-dimensional representation of mean and variance information with minimal distortion.[137] The relationship between random Gaussian directions and principal curvature is less straightforward. Our results show that the principal curvatures $\kappa_{\pm}^{\alpha,\beta}$ in a two-dimensional loss projection are weighted averages of the Hessian elements $(H_\theta)_{ij}$ in the original space, not weighted averages of the principal curvatures κ_i^θ as claimed by Ref. [123]. Similar arguments apply to projections with dimension larger than 2.

4.1.4. *Hessian trace estimates*

Finally, we point to a connection between curvature measures and Hessian trace estimates. A common way of estimating $\mathrm{tr}(H_\theta)$ without explicitly computing all eigenvalues of H_θ is based on Hutchinson's method[138] and random numerical linear algebra.[139,140] The basic idea behind this approach is to (i) use a random vector $z \in \mathbb{R}^N$ with elements z_i that are distributed according to a distribution function with zero mean and unit variance (*e.g.*, a Rademacher distribution with $\mathrm{Pr}\,(z_i = \pm 1) = 1/2$), and (ii) compute $z^\top H_\theta z$, an unbiased estimator of $\mathrm{tr}(H_\theta)$. That is,

$$\mathrm{tr}(H_\theta) = \mathbb{E}[z^\top H_\theta z]. \tag{39}$$

Recall that Eq. (37) shows that the principal curvature of the expected random loss projection, $\bar{\kappa}^{\alpha,\beta}$, is equal to $\mathrm{tr}(H_\theta)$. Instead of estimating $\mathrm{tr}(H_\theta)$ using Hutchinson's method (39), an alternative Hutchinson-type estimate of this quantity is provided by the mean of the expected values of $\kappa_-^{\alpha,\beta}$ and $\kappa_+^{\alpha,\beta}$, see Eq. (37).

4.2. *Extracting curvature information*

We now study two examples that will help build intuitions. In the first example presented in Sec. 4.2.1, we study a critical point θ^* of an N-dimensional loss function $L(\theta)$ for which (i) all principal curvatures have the same magnitude and (ii) the number of positive curvature directions is equal to the number of negative curvature directions. In this example, saddles can be correctly detected by ensemble means but success depends on the averaging process used. In the second example presented in Sec. 4.2.2, we use a loss function associated with an unequal number of negative and positive curvature directions. For the different curvature measures derived in Sec. 4.1.3, we find that random projections cannot identify the underlying saddle point. In Sec. 4.2.3, we will use the two example loss functions to discuss how curvature-based Hessian trace estimates relate to those obtained with the original Hutchinson's method.

4.2.1. *Equal number of curvature directions*

The loss function of our first example is

$$L(\theta) = \frac{1}{2}\theta_{2n+1}\left(\sum_{i=1}^{n}\theta_i^2 - \theta_{i+n}^2\right), \quad n \in \mathbb{Z}_+ , \tag{40}$$

where we set $N = 2n + 1$. A critical point θ^* of the loss function (40) satisfies

$$(\nabla_\theta L)(\theta^*) = \begin{pmatrix} \theta_1^* \theta_{2n+1}^* \\ \vdots \\ \theta_n^* \theta_{2n+1}^* \\ -\theta_{n+1}^* \theta_{2n+1}^* \\ \vdots \\ -\theta_{2n}^* \theta_{2n+1}^* \\ \frac{1}{2}\left(\sum_{i=1}^{n}\theta_i^{*2} - \theta_{i+n}^{*2}\right) \end{pmatrix} = 0 . \tag{41}$$

The Hessian at the critical point $\theta^* = (\theta_1^*, \ldots, \theta_{2n}^*, \theta_{2n+1}^*) = (0, \ldots, 0, 1)$ is

$$H_\theta = \mathrm{diag}(\underbrace{1, \ldots, 1}_{n \text{ times}}, \underbrace{-1, \ldots, -1}_{n \text{ times}}, 0) . \tag{42}$$

Because H_θ has positive and negative eigenvalues, the critical point is a saddle. The corresponding principal curvatures are $\kappa_i^\theta \in \{-1, 1\}$ ($i \in \{1, \ldots, N-1\}$) and $\kappa_N^\theta = 0$. In this example, the mean curvature H, as defined in Eq. (38), is equal to 0. According to Eq. (36), the principal curvature, $\bar{\kappa}^{\alpha,\beta}$, associated with the expected, dimension-reduced Hessian $H_{\alpha,\beta}$ is also equal to 0, erroneously indicating an apparently flat loss landscape if one would use $\bar{\kappa}^{\alpha,\beta}$ as the main measure of curvature. To compare the convergence of different curvature measures as a function of the number of loss projections S, we will now study the ensemble mean

$$\langle X \rangle = \frac{1}{S}\sum_{k=1}^{S} X^{(k)} \tag{43}$$

of different quantities of interest X such as Hessian elements and principal curvature measures in dimension-reduced space. Here, $X^{(k)}$ is the k-th realization (or sample) of X.

We first study the dependence of ensemble means $\langle (H_{\alpha,\beta})_{ij} \rangle$ ($i, j \in \{1, 2\}$) of elements of the dimension-reduced Hessian $H_{\alpha,\beta}$ on the number of samples S. According to Eq. (36), the diagonal elements of $\mathbb{E}[H_{\alpha,\beta}]$ are proportional to the mean curvature of the original high-dimensional space

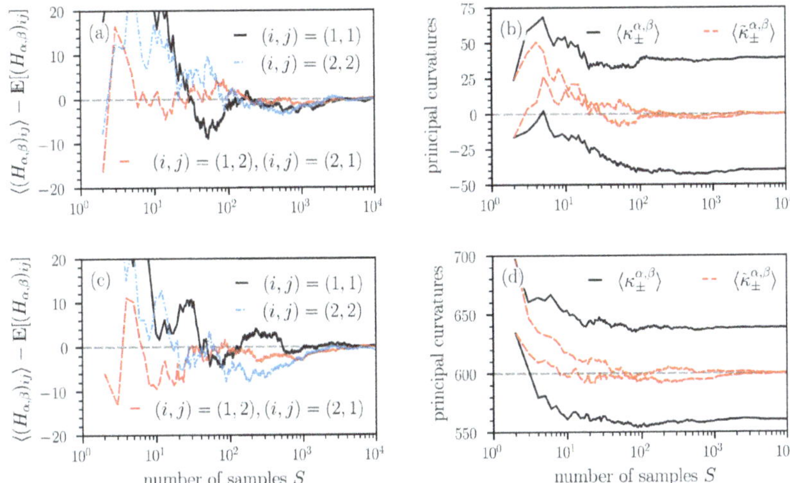

Figure 7. Convergence of the ensemble mean (43) of Hessian elements and curvature measures as a function of the number of random projections S. (a,c) The deviation of the ensemble means $\langle (H_{\alpha,\beta})_{ij} \rangle$ ($i,j \in \{1,2\}$) of Hessian elements from the corresponding expected values as a function of S. Notice that the expected value of the diagonal elements $(H_{\alpha,\beta})_{11}$ and $(H_{\alpha,\beta})_{22}$ is equal to $\bar{\kappa}^{\alpha,\beta}$ (*i.e.*, to the sum of principal curvatures in the original space), see Eqs. (36) and (37). A relatively large number of random projections between 10^3 and 10^4 is required to keep the deviations at values smaller than about 2–4. (b,d) The ensemble means $\langle \kappa_{\pm}^{\alpha,\beta} \rangle$ [see Eq. (35)] and $\langle \tilde{\kappa}_{\pm}^{\alpha,\beta} \rangle$ [see Eq. (44)] as a function of S. Dashed grey lines represent $\bar{\kappa}^{\alpha,\beta} = \mathrm{tr}(H_\theta)$. In panels (a,b) and (c,d), the N-dimensional loss functions are given by Eqs. (40) and (45), respectively. We evaluate the corresponding Hessians (42) and (46) at the saddle point $\theta^* = (\theta_1^*, \ldots, \theta_{2n}^*, \theta_{2n+1}^*) = (0, \ldots, 0, 1)$. In both loss functions, we set $n = 500$ and in loss function (45) we set $\tilde{n} = 800$.

and are thus useful to examine curvature properties of high-dimensional loss functions. Fig. 7(a) shows the convergence of the ensemble means $\langle (H_{\alpha,\beta})_{ij} \rangle$ toward the expected values $\mathbb{E}[(H_{\alpha,\beta})_{ij}]$ as a function of S. Note that $\mathbb{E}[(H_{\alpha,\beta})_{ij}] = 0$ for all i,j. For a few dozen loss projections, the deviations of some of the ensemble means from the corresponding expected values reach values larger than 20. A relatively large number of loss projections S between 10^3–10^4 is required to keep these deviations at values that are smaller than about 2–4. The solid black and red lines in Fig. 7(b), respectively, show the ensemble means $\langle \kappa_{\pm}^{\alpha,\beta} \rangle$ and

$$\langle \tilde{\kappa}_{\pm}^{\alpha,\beta} \rangle = \frac{1}{2} \left(\langle A \rangle + \langle C \rangle \pm \sqrt{4\langle B \rangle^2 + (\langle A \rangle - \langle C \rangle)^2} \right) \qquad (44)$$

as a function of S. Since $\mathbb{E}[A] = \mathbb{E}[C] = (H_\theta)^i{}_i$ and $\mathbb{E}[B] = 0$ [see Eq. (36)],

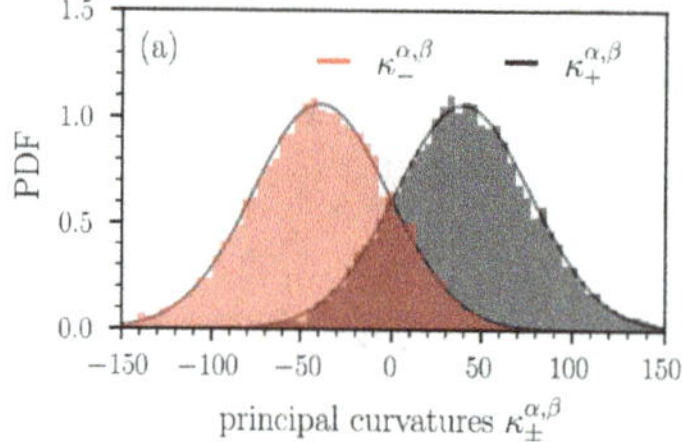 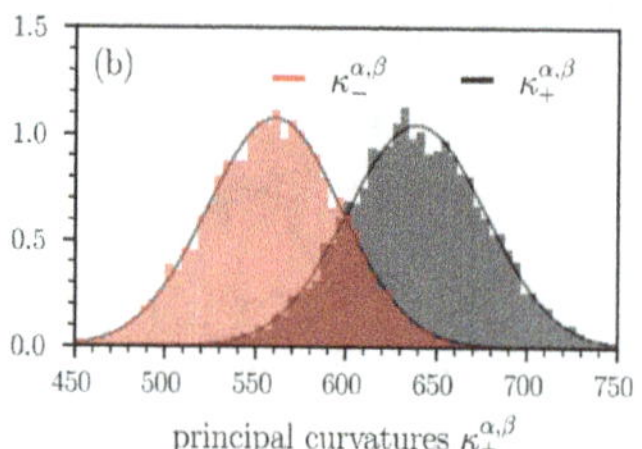

Figure 8.   Distribution of principal curvatures $\kappa_-^{\alpha,\beta}$ (red bars) and $\kappa_+^{\alpha,\beta}$ (black bars). In panels (a) and (b), the loss functions are given by Eqs. (40) and (45), respectively. We evaluate the corresponding Hessians (42) and (46) at the saddle point $\theta^* = (\theta_1^*,\ldots,\theta_{2n}^*,\theta_{2n+1}^*) = (0,\ldots,0,1)$. In both loss functions, we set $n = 500$ and in loss function (45) we set $\tilde{n} = 800$. In panel (a), the probability $\Pr(\kappa_+^{\alpha,\beta}\kappa_-^{\alpha,\beta} > 0)$ that the critical point in the lower-dimensional, random projection does not appear as a saddle is about 0.3, and it is 1 in panel (b). Histograms are based on 10,000 random projections used to compute $\kappa_\pm^{\alpha,\beta}$. Solid grey lines are Gaussian approximations of the empirical distributions.

we have that $\langle \tilde{\kappa}_\pm^{\alpha,\beta} \rangle = \bar{\kappa}^{\alpha,\beta}$ in the limit $S \to \infty$. In the current example, the ensemble means $\langle \tilde{\kappa}_\pm^{\alpha,\beta} \rangle$ thus approach $\bar{\kappa}^{\alpha,\beta} = 0$ for large numbers of samples S, represented by the dashed red lines in Fig. 7(b). The ensemble means $\langle \kappa_\pm^{\alpha,\beta} \rangle$ converge towards values of opposite sign, indicating a saddle point.

For a sample size of $S = 10^4$, we show the distribution of the principal curvatures $\kappa_\pm^{\alpha,\beta}$ in Fig. 8(a). We observe that the distributions are plausibly Gaussian. We also calculate the probability $\Pr(\kappa_+^{\alpha,\beta}\kappa_-^{\alpha,\beta} > 0)$ that the critical point in the lower-dimensional, random projection does not appear as a saddle (*i.e.*, $\kappa_+^{\alpha,\beta}\kappa_-^{\alpha,\beta} > 0$). For the example shown in Fig. 8(a), we find that $\Pr(\kappa_+^{\alpha,\beta}\kappa_-^{\alpha,\beta} > 0) \approx 0.3$. That is, in about 30% of the simulated projections, the lower-dimensional loss landscape wrongly indicates that it does not correspond to a saddle.

Our first example, which is based on the loss function (40), shows that the principal curvatures in the lower-dimensional representation of $L(\theta)$ may capture the saddle behavior in the original space if one computes ensemble means $\langle \kappa_\pm^{\alpha,\beta} \rangle$ in the lower-dimensional space, see Fig. 7(b). However, if one first calculates ensemble means of the elements of the dimension-reduced Hessian $H_{\alpha,\beta}$ to infer $\langle \tilde{\kappa}_\pm^{\alpha,\beta} \rangle$, the loss landscape appears to be flat in this example. We thus conclude that different ways of computing ensemble means (either before or after calculating the principal curvatures) may lead to different results with respect to the "flatness" of a dimension-reduced loss landscape.

4.2.2. *Unequal number of curvature directions*

In the next example, we will show that random projections cannot identify certain saddle points regardless of the underlying averaging process. We consider the loss function

$$L(\theta) = \frac{1}{2}\theta_{2n+1}\left(\sum_{i=1}^{\tilde{n}}\theta_i^2 - \sum_{i=\tilde{n}+1}^{2n}\theta_i^2\right), \quad n \in \mathbb{Z}_+, n < \tilde{n} \leq 2n, \qquad (45)$$

where we use the convention $\sum_{i=a}^{b}(\cdot) = 0$ if $a > b$. The Hessian at the critical point $(\theta_1^*, \ldots, \theta_{2n}^*, \theta_{2n+1}^*) = (0, \ldots, 0, 1)$ is

$$H_\theta = \mathrm{diag}(\underbrace{1, \ldots, 1}_{\tilde{n} \text{ times}}, \underbrace{-1, \ldots, -1}_{2n-\tilde{n} \text{ times}}, 0). \qquad (46)$$

As in the previous example, the critical point is again a saddle, but the mean curvature is $H = 2(\tilde{n} - n)/N > 0$. In the following numerical experiments, we set $n = 500$ and $\tilde{n} = 800$. Fig. 7(c) shows that the ensemble means $\langle (H_{\alpha,\beta})_{ij} \rangle$ converge towards the expected value $\mathbb{E}[(H_{\alpha,\beta})_{ij}]$ as the number of samples increases. We again observe that a relatively large number of random loss projections S between 10^3 and 10^4 is required to keep the deviations of ensemble means from their corresponding expected values small. Because of the dominance of positive principal curvatures κ_i^θ in the original space, the corresponding ensemble means of principal curvatures (*i.e.*, $\langle \kappa_\pm^{\alpha,\beta} \rangle$, $\langle \tilde{\kappa}_\pm^{\alpha,\beta} \rangle$) in the lower-dimensional representation approach positive values [see Fig. 7(d)]. The distribution of $\kappa_\pm^{\alpha,\beta}$ indicates that the probability of observing a saddle in the lower-dimensional loss landscape is vanishingly small, see Fig. 8(b). In this second example, both ways of computing ensemble means, before and after calculating the lower-dimensional principal curvatures, mistakenly suggest that the saddle in the original space is a minimum in dimension-reduced space.

To summarize, for both loss functions (40) and (45), the saddle point $\theta^* = (0, \ldots, 0, 1)$ in the original loss function $L(\theta)$ is often misrepresented in lower-dimensional representations $L(\theta + \alpha\eta + \beta\delta)$ if random directions are used. Depending on (i) the employed curvature measure and (ii) the index of the underlying Hessian H_θ in the original space, the saddle $\theta^* = (0, \ldots, 0, 1)$ may appear erroneously as either a minimum, maximum, or an almost flat region.

If the critical point were a minimum or maximum (*i.e.*, a critical point associated with a positive definite or negative definite Hessian H_θ), it would

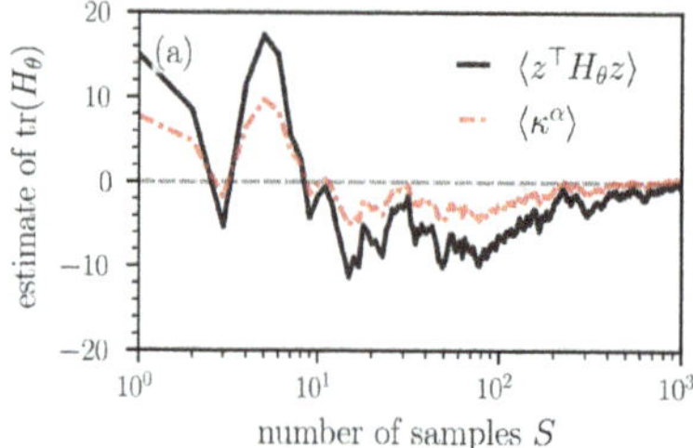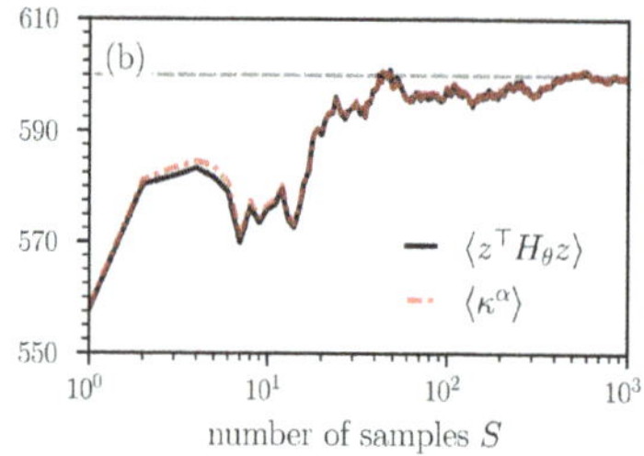

Figure 9.   Estimating the trace of the Hessian H_θ. In panels (a) and (b), the loss functions are given by Eqs. (40) and (45), respectively. We evaluate the corresponding Hessians (42) and (46) at the saddle point $\theta^* = (\theta_1^*, \ldots, \theta_{2n}^*, \theta_{2n+1}^*) = (0, \ldots, 0, 1)$. In both loss functions, we set $n = 500$ and in loss function (45) we set $\tilde{n} = 800$. Solid black and dash-dotted red lines represent Hutchinson $[\langle z^\top H_\theta z\rangle$; see Eq. (39)] and curvature-based $(\langle \kappa^\alpha\rangle)$ estimates of $\mathrm{tr}(H_\theta)$, respectively. We compute ensemble means $\langle\cdot\rangle$ as defined in Eq. (43) for different numbers of random projections S. The trace estimates in panels (a) and (b), respectively, converge towards the true trace values $\mathrm{tr}(H_\theta) = 0$ and $\mathrm{tr}(H_\theta) = 600$ that are indicated by dashed grey lines. In both methods, the same random vectors with elements that are distributed according to a standard normal distribution are used. For the curvature-based estimation of $\mathrm{tr}(H_\theta)$, we perform least-square fits of $L(\theta^* + \alpha\eta)$ over an interval $\alpha \in [-0.05, 0.05]$.

be correctly represented in the corresponding expected random projection because the sign of its principal curvature $\bar{\kappa}^{\alpha,\beta}$, which is proportional to the sum of all eigenvalues κ_i^θ of the Hessian H_θ in the original loss space, would be equal to the sign of the principal curvatures κ_i^θ. However, such points are scarce in high-dimensional loss spaces.[141,142]

Finally, because of the confounding factors associated with the inability of random projections to correctly identify saddle information, it does not appear advisable to use "flatness" around a critical point in a lower-dimensional random loss projection as a measure of generalization error.[123]

4.2.3.   *Curvature-based Hessian trace estimation*

In accordance with Sec. 4.1.4, we now use the loss functions (40) and (45) to compare the convergence behavior between the original Hutchinson method (39) and the curvature-based trace estimation (37). Instead of two random directions, we use one random Gaussian direction η and perform least-squares fits for 50 equidistant values of α in the interval $[-0.05, 0.05]$ to extract estimates of $\mathrm{tr}(H_\theta)$ from $L(\theta^* + \alpha\eta)$. We use the same random Gaussian directions in Hutchinson's method.

Fig. 9 shows how the Hutchinson and curvature-based trace estimates converge towards the true trace values, 0 for the loss function (40) and 600 for the loss function (45) with $n = 500$ and $\tilde{n} = 800$. Given that we use the same random vectors in both methods, their convergence behavior towards the true trace value is similar.

With the curvature-based method, one can produce Hutchinson-type trace estimates without computing Hessian-vector products. However, it requires the user to specify an appropriate interval for α so that the mean curvature can be properly estimated in a quadratic-approximation regime. It also requires a sufficiently large number of points in this interval. Therefore, it may be less accurate than the original Hutchinson method.

4.3. *Hessian directions*

Given the described shortcomings of random projections incorrectly identifying saddles of high-dimensional loss functions, we suggest to use Hessian directions (*i.e.*, the eigenbasis of H_θ) as directions η, δ in $L(\theta^* + \alpha\eta + \beta\delta)$.

For Eq. (45) with $n = 900, \tilde{n} = 1000$, we show projections along different Hessian directions in Fig. 10(a–c). We observe that different Hessian directions indicate different types of critical points in dimension-reduced space. If the eigenvalues associated with the Hessian directions η, δ have different signs, the corresponding lower-dimensional loss landscape is a saddle, see Fig. 10(a). If both eigenvalues have the same sign, the loss landscape is either a minimum [see Fig. 10(b): both signs are positive] or it is a maximum [see Fig. 10(c): both signs are negative]. If one uses a random projection instead, the resulting lower-dimensional loss landscape often appears to be a minimum in this example, see Fig. 10(d). To quantify the proportion of random projections that correctly identify the saddle with $n = 900, \tilde{n} = 1000$, we generated 10,000 realizations of $\kappa_\pm^{\alpha,\beta}$, see Eq. (35). We find that the signs of $\kappa_\pm^{\alpha,\beta}$ were different in only about 0.5% of all simulated realizations. That is, in this example the principal curvatures $\kappa_\pm^{\alpha,\beta}$ indicate a saddle in only about 0.5% of the studied projections.

In accordance with related works that use Hessian information,[143,144] our proposal uses *Hessian-vector products* (HVPs). We first compute the largest-magnitude eigenvalue and then use an annihilation method[145] to compute the second largest-magnitude eigenvalue of opposite sign, see Algorithm 1. The corresponding eigenvectors are the dominant Hessian directions. Other deflation techniques can be used to find additional Hessian directions.

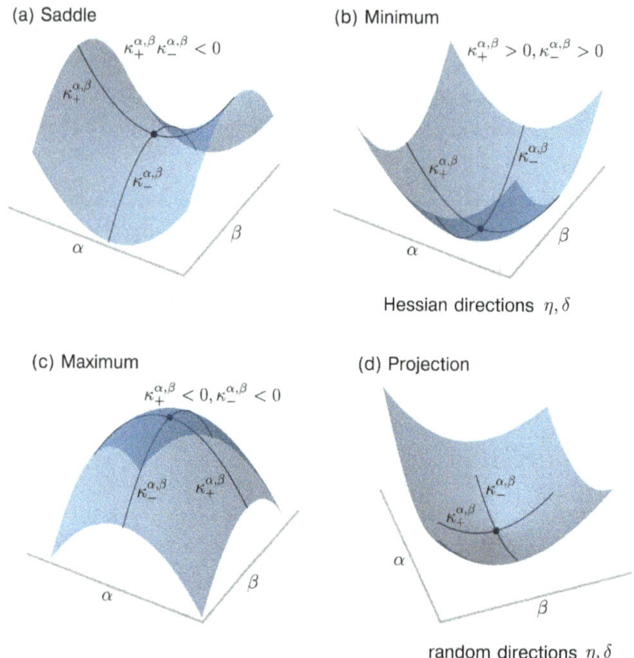

Figure 10. Dimensionality-reduced loss $L(\theta^* + \alpha\eta + \beta\delta)$ of Eq. (45) with $n = 900, \tilde{n} = 1000$ for different directions η, δ. (a–c) The directions η, δ correspond to eigenvectors of the Hessian H_θ of Eq. (45). If the eigenvalues associated with η, δ have different signs, the corresponding loss landscape is a saddle as depicted in panel (a). If the eigenvalues associated with η, δ have the same sign, the corresponding loss landscape is either a minimum (both signs are positive) as shown in panel (b) or a maximum (both signs are negative) as shown in panel (c). Because there is an excess of $\tilde{n} - n = 100$ positive eigenvalues in H_θ, a projection onto a dimension-reduced space that is spanned by the random directions η, δ is often associated with an apparently convex loss landscape. An example of such an apparent minimum is shown in panel (d).

We employ HVPs to compute Hessian directions without an explicit representation of H_θ using the identity

$$\nabla_\theta[(\nabla_\theta L)^\top v] = (\nabla_\theta \nabla_\theta L)v + (\nabla_\theta L)^\top \nabla_\theta v = H_\theta v. \qquad (47)$$

In the first step, the gradient $(\nabla_\theta L)^\top$ is computed using reverse-mode autodifferentiation (AD) to then compute the scalar product $[(\nabla_\theta L)^\top v]$. In the second step, we again apply reverse-mode AD to the computational graph associated with the scalar product $[(\nabla_\theta L)^\top v]$. Because the vector v does not depend on θ (*i.e.*, $\nabla_\theta v = 0$), the result is $H_\theta v$. One may also use forward-mode AD in the second step to provide a more memory efficient implementation.

Algorithm 1 Compute dominant Hessian directions

1: $L_1 = \text{LinearOperator}((N, N),\ \text{matvec} = \text{hvp})$

 initialize linear operator for HVP calculation

2: eigval1, eigvec1 $= \text{solve\_lm\_evp}(L_1)$

 compute largest-magnitude eigenvalue and corresponding eigenvector associated with operator L_1

3: shifted_hvp(vec) $=$ hvp(vec) - eigval1*vec

 define shifted HVP

4: $L_2 = \text{LinearOperator}((N, N),\ \text{matvec} = \text{shifted\_hvp})$

 initialize linear operator for shifted HVP calculation

5: eigval2, eigvec2 $= \text{solve\_lm\_evp}(L_2)$

 compute largest-magnitude eigenvalue and corresponding eigenvector associated with operator L_2

6: eigval2 $+=$ eigval1

7: **if** eigval1 $>= 0$ **then**

8: maxeigval, maxeigvec $=$ eigval1, eigvec1

9: mineigval, mineigvec $=$ eigval2, eigvec2

10: **else**

11: maxeigval, maxeigvec $=$ eigval2, eigvec2

12: mineigval, mineigvec $=$ eigval1, eigvec1

13: **end if**

14: **return:** maxeigval, maxeigvec, mineigval, mineigvec

4.3.1. *Image classification*

As an illustration of a loss landscape of an ANN employed in image classification, we focus on the ResNet-56 architecture that has been trained as detailed in Ref. [123] on CIFAR-10 using stochastic gradient descent (SGD) with Nesterov momentum. The number of parameters of this ANN is 855,770. The training and test losses at the local optimum found by SGD are 9.20×10^{-4} and 0.29, respectively; the corresponding accuracies are 100.00 and 93.66, respectively.

In accordance with Ref. [123], we apply filter normalization to random directions. This and related normalization methods are often employed when generating random projections. One reason is that simply adding random vectors to parameters of a neural network loss function (or parameters of other functions) does not consider the range of parameters associated with different elements of that function. As a result, random

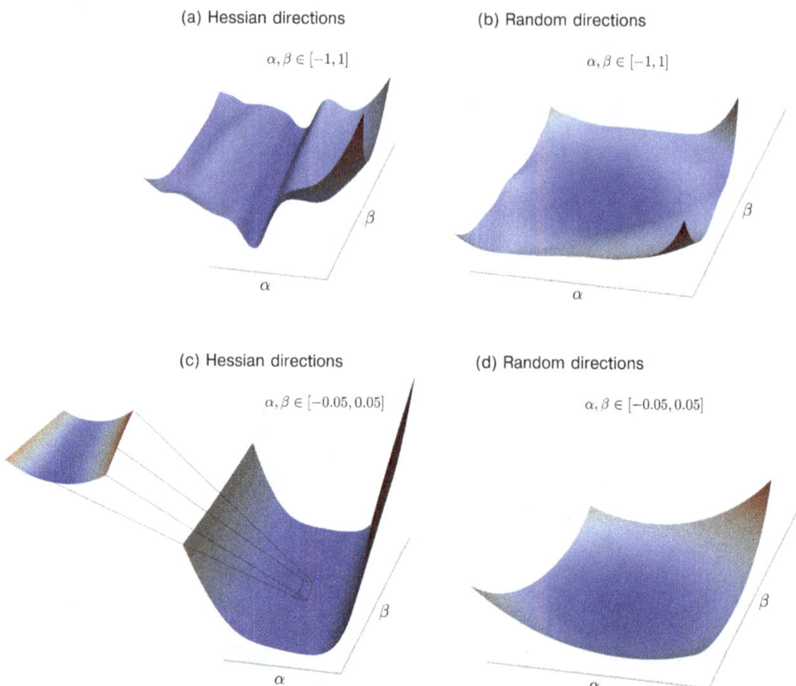

Figure 11.   Loss landscape projections for ResNet-56. (a,c) The projection directions η, δ are given by the eigenvectors associated with the largest and smallest eigenvalues of the Hessian H_θ, respectively. The zoomed inset in panel (c) shows the loss landscape for $(\alpha, \beta) \in [-0.01, 0.005] \times [-0.05, 0.05]$. We observe a decreasing loss along the negative β-axis. (b,d) The projection directions η, δ are given by random vectors. The domains of (α, β) in panels (a,b) and (c,d) are $[-1, 1] \times [-1, 1]$ and $[-0.05, 0.05] \times [-0.05, 0.05]$, respectively. All shown cross-entropy loss landscapes are based on evaluating the CIFAR-10 training dataset that consists of 50,000 images.

perturbations may be too small or large to properly resolve the influence of certain parameters on a given function.

When calculating Hessian directions, we are directly taking into account the parameterization of the underlying functions that we want to visualize. Therefore, there is no need for an additional rescaling of different parts of the perturbation vector. Still, reparameterizations of a neural network can result in changes of curvature properties, see, *e.g.*, Theorem 4 in Ref. [121].

Fig. 11 shows the two-dimensional projections of the loss function (cross entropy loss) around a local critical point. The smallest and largest eigenvalues are -16.4 and 5007.9, respectively. This means that the found critical point is a saddle with a maximum negative curvature that is more than two orders of magnitude smaller than the maximum positive curvature at that point. The saddle point is clearly visible in Fig. 11(a,c). We observe in the zoomed inset in Fig. 11(c) that the loss decreases along the negative β-axis.

If random directions are used, the corresponding projections indicate that the optimizer converged to a local minimum and not to a saddle point, see Fig. 11(b,d). Overall, the ResNet-56 visualizations that we show in Fig. 11 exhibit structural similarities to those that we generated using the simple loss model (45), see Fig. 10(a,d).

4.3.2. *Function approximation*

As another example, we compare loss function visualizations that are based on random and Hessian directions in a function-approximation task. In accordance with Ref. [146], we consider the smooth one-dimensional function

$$f(x) = \log(\sin(10x) + 2) + \sin(x), \tag{48}$$

where $x \in [-1, 1)$. To approximate $f(x)$, we use two fully connected neural networks (FCNNs) with 2 and 10 layers, respectively. Each layer has 100 ReLU activations. The numbers of parameters of the 2 and 10 layer architectures are 20,501 and 101,301, respectively. The training data is based on 50 points that are sampled uniformly at random from the interval $[-1, 1)$. We train both ANNs using a mean squared error (MSE) loss function and SGD with a learning rate of 0.1. The 2 and 10 layer architectures are respectively trained for 100,000 and 50,000 epochs to reach loss values of less than 10^{-4}. The best model was saved and evaluated by calculating the MSE loss for 1,000 points that were sampled uniformly at random from the interval $[-1, 1)$.[1]

Fig. 12 shows heatmaps of the training and test loss landscapes of both ANNs along random and dominant Hessian directions. Black crosses in Fig. 12 indicate loss minima. In line with the previous example, which focused on an image classification ANN, we observe for both FCNNs that random projections are associated with loss values that increase along both directions δ, η and for both training and test data, see Fig. 12(a,b,e,f). For these projections, we find that the loss minima are very close to or at the origin of the loss space. The situation is different in the loss projections that

---

[1] We present an animation illustrating the evolution of random and Hessian loss projections during training at `https://vimeo.com/787174225`.

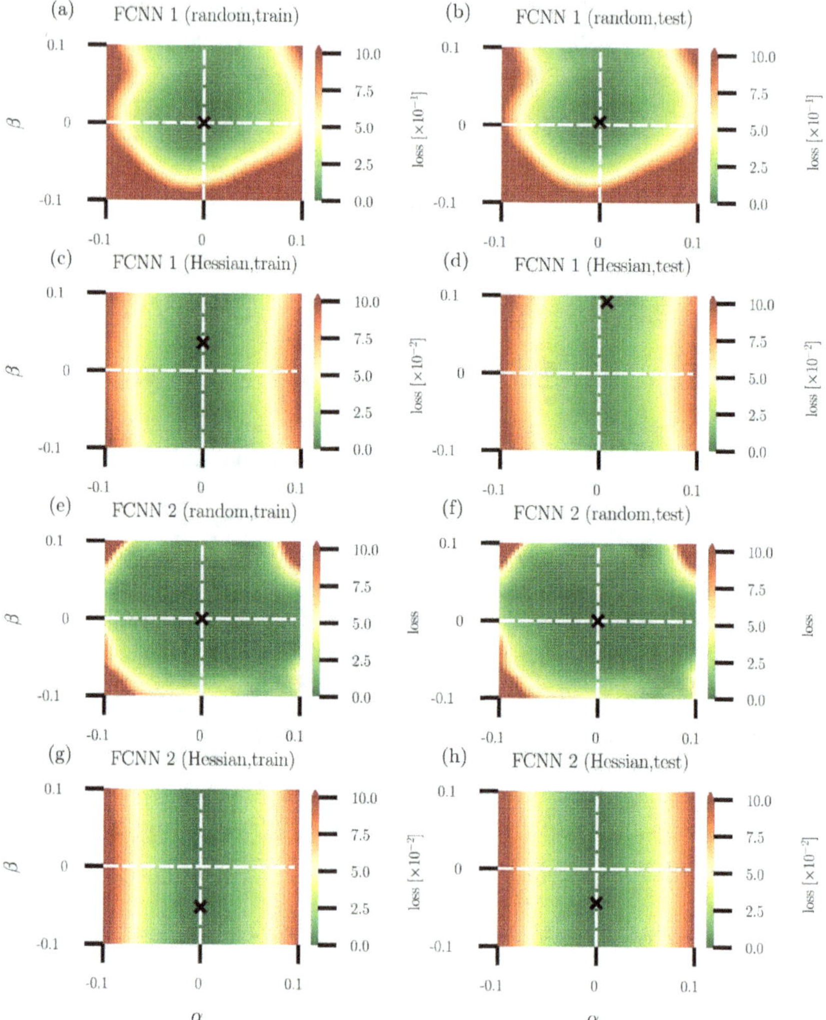

Figure 12. Heatmaps of the mean squared error (MSE) loss along random and dominant Hessian directions for a function-approximation task. (a–d) Loss heatmaps for a fully connected neural network (FCNN) with 2 layers and 100 ReLU activations per layer [(a,c): training data, (b,d): test data]. (e–h) Loss heatmaps for an FCNN with 10 layers and 100 ReLU activations per layer [(e,g): training data, (f,h): test data]. Random directions are used in panels (a,b,e,f) while Hessian directions are used in panels (c,d,g,h). Green and red regions indicate small and large mean squared error (MSE) loss values, respectively. Black crosses indicate the positions of loss minima in the shown domain. Both neural networks, FCNN 1 and FCNN 2, are trained to approximate the smooth one-dimensional function (48).

are based on Hessian directions. Fig. 12(c) shows that the loss minimum for the 2-layer FCNN is not located at the origin but at $(\alpha, \beta) \approx (0, 0.04)$.

We find that the value of the training loss at that point is more than 9% smaller than the smallest training loss found in a random projection plot. The corresponding test loss is about 1% smaller than the test loss associated with the smallest training loss in the random projection plot. For the 10-layer FCNN, the smallest training loss in the Hessian direction projection is more than 26% smaller than the smallest training loss in the randomly projected loss landscape, see Fig. 12(e,g). The corresponding test loss in the Hessian direction plot is about 6% smaller than the corresponding test loss minimum in the random direction plot, see Fig. 12(f,h). Notice that both the training and test loss minima in the random direction heatmaps in Fig. 12(e,f) are located at the origin while they are located at $(\alpha, \beta) \approx (0, -0.05)$ in the Hessian direction heatmaps in Fig. 12(g,h).

These results show that it is possible to improve the training loss values along the dominant negative curvature direction. Smaller loss values may also be achievable with further gradient-based training iterations, so one has to consider tradeoffs between gradient and curvature-based optimization methods.

5. Conclusions

Over several decades, research at the intersection of statistical mechanics, neuroscience, and computer science has significantly enhanced our understanding of information processing in both living systems and machines. With the increasing computing power, numerous learning algorithms conceived in the latter half of the 20th century have found widespread application in tasks such as image classification, natural language processing, and anomaly detection.

The Ising model, one of the most thoroughly studied models in statistical mechanics, shares close connections with ANNs such as Hopfield networks and Boltzmann machines. In the first part of this chapter, we described these connections and provided an illustrative example of a Boltzmann machine that learned to generate new Ising configurations based on a set of training samples.

Despite decades of active research in artificial intelligence and machine learning, providing mechanistic insights into the representational capacity and learning dynamics of modern ANNs remains challenging due to their high dimensionality and complex, non-linear structure. Visualizing the loss landscapes associated with ANNs poses a particular challenge, as one must appropriately reduce dimensionality to visualize landscapes in two or three dimensions. To put it in the words of Ker-Chau Li, "There are too many directions to project a high-dimensional data set and unguided plotting can be time-consuming and fruitless". One approach is to employ random projections. However, this method often inaccurately depicts saddle points in high-dimensional loss landscapes as apparent minima in low-dimensional random projections.

In the second part of this chapter, we discussed various curvature measures that can be employed to characterize critical points in high-dimensional loss landscapes. Additionally, we outlined how Hessian directions (*i.e.*, the eigenvectors of the Hessian at a critical point) can inform a more structured approach to projecting high-dimensional functions onto lower-dimensional representations.

While much of contemporary research in machine learning is driven by practical applications, incorporating theoretical contributions rooted in concepts from statistical mechanics and related disciplines holds promise for continuing to guide the development of learning algorithms.

Acknowledgements

LB acknowledges financial support from hessian.AI and the ARO through grant W911NF-23-1-0129.

References

1. L. Zdeborová and F. Krzakala, Statistical physics of inference: Thresholds and algorithms, *Advances in Physics*. **65**(5), 453–552 (2016).
2. G. Carleo, I. Cirac, K. Cranmer, L. Daudet, M. Schuld, N. Tishby, L. Vogt-Maranto, and L. Zdeborová, Machine learning and the physical sciences, *Reviews of Modern Physics*. **91**(4), 045002 (2019).
3. P. Mehta, M. Bukov, C.-H. Wang, A. G. Day, C. Richardson, C. K. Fisher, and D. J. Schwab, A high-bias, low-variance introduction to machine learning for physicists, *Physics Reports*. **810**, 1–124 (2019).
4. D. J. Amit, H. Gutfreund, and H. Sompolinsky, Spin-glass models of neural networks, *Physical Review A*. **32**(2), 1007 (1985).
5. T. Ising, R. Folk, R. Kenna, B. Berche, and Y. Holovatch, The fate of Ernst Ising and the fate of his model, *Journal of Physical Studies*. **21**, 3002 (2017).
6. R. Folk. The survival of Ernst Ising and the struggle to solve his model. In ed. Y. Holovatch, *Order, Disorder and Criticality: Advanced Problems of Phase Transition Theory*, pp. 1–77. World Scientific, Singapore (2023).

7. A. Rosenblueth, N. Wiener, and J. Bigelow, Behavior, purpose and teleology, *Philosophy of Science.* **10**(1), 18–24 (1943).

8. N. Wiener, *Cybernetics or Control and Communication in the Animal and the Machine.* MIT Press, Boston, MA, USA (1948).

9. E. Thorndike, *Fundamentals of Learning.* Teachers College, Columbia University, New York City, NY, USA (1932).

10. I. S. Berkeley, The curious case of connectionism, *Open Philosophy.* **2**(1), 190–205 (2019).

11. M. Minsky and S. Papert, *Perceptrons: An Introduction to Computational Geometry,* 2nd edn. MIT Press, Cambridge, MA, USA (1972).

12. S. Russell and P. Norvig, *Artificial Intelligence: A Modern Approach,* 3rd edn. Prentice Hall, Upper Saddle River, NJ, USA (2009).

13. V. Vapnik and A. Chervonenkis, *Theory of Pattern Recognition: Statistical Learning Problems.* Nauka, Moscow, Russia (1974). In Russian.

14. T. M. Mitchell, *Machine Learning.* McGraw-Hill, New York City, NY, USA (1997).

15. W. S. McCulloch and W. Pitts, A logical calculus of the ideas immanent in nervous activity, *The Bulletin of Mathematical Biophysics.* **5**, 115–133 (1943).

16. T. H. Abraham, Nicolas Rashevsky's mathematical biophysics, *Journal of the History of Biology.* **37**(2), 333–385 (2004).

17. A. Turing, On computable numbers with an application to the Entscheidungsproblem, *Proceedings of the London Mathematical Society.* **58**, 230–265 (1937).

18. S. C. Kleene, *Representation of Events in Nerve Nets and Finite Automata,* In eds. C. E. Shannon and J. McCarthy, *Automata Studies. (AM-34),* vol. 34, pp. 3–42. Princeton University Press, Princeton, MA, USA (1956).

19. D. O. Hebb, *The organization of behavior: A neuropsychological theory.* Wiley & Sons, New York City, NY, USA (1949).

20. S. Löwel and W. Singer, Selection of intrinsic horizontal connections in the visual cortex by correlated neuronal activity, *Science.* **255**(5041), 209–212 (1992).

21. F. Rosenblatt, The perceptron: a probabilistic model for information storage and organization in the brain, *Psychological Review.* **65**(6), 386 (1958).

22. F. Rosenblatt et al., *Principles of neurodynamics: Perceptrons and the theory of brain mechanisms.* vol. 55, Spartan Books, Washington, DC, USA (1962).

23. B. Widrow and M. Hoff. Adaptive switching circuits. In *1960 IRE WESCON Convention Record,* pp. 96–104 (1960).

24. S. Amari, A theory of adaptive pattern classifiers, *IEEE Transactions on Electronic Computers.* **EC-16**(3), 299–307 (1967).

25. A. G. Ivakhnenko and V. G. Lapa, *Cybernetics and Forecasting Techniques.* American Elsevier Publishing Company, Philadelphia, PA, USA (1967).

26. A. G. Ivakhnenko, Polynomial theory of complex systems, *IEEE Transactions on Systems, Man, and Cybernetics.* (4), 364–378 (1971).

27. H. J. Kelley, Gradient theory of optimal flight paths, *Ars Journal.* **30**(10),

947–954 (1960).

28. S. Linnainmaa. The representation of the cumulative rounding error of an algorithm as a taylor expansion of the local rounding errors. Master's thesis, University of Helsinki (1970).

29. S. Linnainmaa, Taylor expansion of the accumulated rounding error, *BIT Numerical Mathematics.* **16**(2), 146–160 (1976).

30. P. J. Werbos. Applications of advances in nonlinear sensitivity analysis. In eds. R. F. Drenick and F. Kozin, *System Modeling and Optimization*, pp. 762–770, Springer Berlin Heidelberg, Berlin, Heidelberg, DE (1982).

31. D. E. Rumelhart, G. E. Hinton, and R. J. Williams, Learning representations by back-propagating errors, *Nature.* **323**(6088), 533–536 (1986).

32. A. J. Noest. Phasor neural networks. In ed. D. Z. Anderson, *Neural Information Processing Systems, Denver, Colorado, USA, 1987*, pp. 584–591, American Institue of Physics (1987).

33. A. J. Noest, Discrete-state phasor neural networks, *Physical Review A.* **38**(4), 2196 (1988).

34. A. J. Noest, Associative memory in sparse phasor neural networks, *Europhysics Letters.* **6**(5), 469 (1988).

35. M. Kobayashi, Exceptional reducibility of complex-valued neural networks, *IEEE Transactions on Neural Networks.* **21**(7), 1060–1072 (2010).

36. T. Minemoto, T. Isokawa, H. Nishimura, and N. Matsui, Quaternionic multistate Hopfield neural network with extended projection rule, *Artificial Life and Robotics.* **21**, 106–111 (2016).

37. H. Zhang, M. Gu, X. Jiang, J. Thompson, H. Cai, S. Paesani, R. Santagati, A. Laing, Y. Zhang, M. Yung, et al., An optical neural chip for implementing complex-valued neural network, *Nature Communications.* **12**(1), 457 (2021).

38. L. Böttcher and M. A. Porter, Complex networks with complex weights, *Physical Review E.* **109**(2), 024314 (2024).

39. T. Bayes, An essay towards solving a problem in the doctrine of chances, *Philosophical Transactions of the Royal Society of London.* **53**, 370–418 (1763).

40. P. Laplace, M'emoire sur la probabilité des causes par les Événements, *M'emoires de Mathématique et de Physique Presentés á l/Académie Royale des Sciences, Par Divers Davans, Lûs dans ses Assemblées.* **6**, 621–656 (1774).

41. T. Seidenfeld, R. A. Fisher's fiducial argument and Bayes' theorem, *Statistical Science.* **7**, 358–368 (1992).

42. T. Seidenfeld. Forbidden fruit: When Epistemic Probability may not take a bite of the Bayesian apple. In eds. W. Harper and G. Wheeler, *Probability and Inference: Essays in Honor of Henry E. Kyburg, Jr.* King's College Publications, London (2007).

43. H. Jeffrey, *The Theory of Probability*, 3rd edn. Oxford Classic Texts in the Physical Sciences, Oxford University Press, Oxford, UK (1939).

44. C. E. Shannon, A mathematical theory of communication, *The Bell System Technical Journal.* **27**(3), 379–423 (1948).

45. E. T. Jaynes, Information theory and statistical mechanics, *Physical Review.*

106(4), 620 (1957).

46. E. T. Jaynes, Information theory and statistical mechanics. II, *Physical Review.* **108**(2), 171 (1957).

47. E. T. Jaynes, The well-posed problem, *Foundations of Physics.* **3**(4), 477–493 (1973).

48. E. T. Jaynes, *Probability Theory: The Logic of Science.* Cambridge University Press, Cambridge, UK (2003).

49. T. Seidenfeld, Why I am not an objective Bayesian., *Theory and Decision.* **11**(4), 413–440 (1979).

50. G. Wheeler, Objective Bayesianism and the problem of non-convex evidence, *The British Journal for the Philosophy of Science.* **63**(3), 841–50 (2012).

51. I. Hacking. Let's not talk about objectivity. In eds. F. Padovani, A. Richardson, and J. Y. Tsou, *Objectivity in Science*, vol. 310, *Boston Studies in the Philosophy and History of Science*, pp. 19–33. Springer, Cham, CH (2015).

52. A. Gelman, Beyond subjective and objective in statistics, *Journal of the Royal Statistical Society. Series A.* **180**(4), 967–1033 (2017).

53. R. Haenni, J.-W. Romeijn, G. Wheeler, and J. Williamson. Logical relations in a statistical problem. In eds. B. Löwe, E. Pacuit, and J.-W. Romeijn, *Foundations of the Formal Sciences VI: Reasoning about probabilities and probabilistic reasoning*, vol. 16, *Studies in Logic*, pp. 49–79, College Publications, London, UK (2009).

54. G. de Cooman, Belief models: An order-theoretic investigation, *Annals of Mathematics and Artificial Intelligence.* **45**(5–34) (2005).

55. R. Haenni, J.-W. Romeijn, G. Wheeler, and J. Williamson, *Probabilistic Logics and Probabilistic Networks.* Synthese Library, Springer, Dordrecht (2011).

56. D. R. Cox, The regression analysis of binary sequences, *Journal of the Royal Statistical Society.* **20**(2), 215–242 (1958).

57. G. E. Hinton and D. van Camp. Keeping the neural networks simple by minimizing the description length of the weights. In ed. L. Pitt, *Proceedings of the Sixth Annual ACM Conference on Computational Learning Theory, COLT 1993, Santa Cruz, CA, USA, July 26-28, 1993*, pp. 5–13, ACM (1993).

58. R. M. Neal, *Bayesian learning for neural networks.* vol. 118, *Lecture Notes in Statistics*, Springer Verlog, New York City, NY, USA (1996).

59. S.-I. Amari, Learning patterns and pattern sequences by self-organizing nets of threshold elements, *IEEE Transactions on Computers.* **100**(11), 1197–1206 (1972).

60. J. J. Hopfield, Neural networks and physical systems with emergent collective computational abilities, *Proceedings of the National Academy of Sciences of the United States of America.* **79**(8), 2554–2558 (1982).

61. T. Isokawa, H. Nishimura, N. Kamiura, and N. Matsui, Associative memory in quaternionic Hopfield neural network, *International Journal of Neural Systems.* **18**(02), 135–145 (2008).

62. H. Ramsauer, B. Schäfl, J. Lehner, P. Seidl, M. Widrich, L. Gruber, M. Holzleitner, T. Adler, D. P. Kreil, M. K. Kopp, G. Klambauer, J. Brandstetter,

and S. Hochreiter. Hopfield networks is all you need. In *9th International Conference on Learning Representations, ICLR 2021, Virtual Event, Austria, May 3-7, 2021* (2021).

63. M. Benedetti, E. Ventura, E. Marinari, G. Ruocco, and F. Zamponi, Supervised perceptron learning vs unsupervised Hebbian unlearning: Approaching optimal memory retrieval in Hopfield-like networks, *The Journal of Chemical Physics.* **156**(10) (2022).

64. L. Böttcher and H. J. Herrmann, *Computational Statistical Physics.* Cambridge University Press, Cambridge, UK (2021).

65. S. E. Fahlman, G. E. Hinton, and T. J. Sejnowski. Massively Parallel Architectures for AI: NETL, Thistle, and Boltzmann Machines. In ed. M. R. Genesereth, *Proceedings of the National Conference on Artificial Intelligence, Washington, DC, USA, August 22-26, 1983*, pp. 109–113, AAAI Press (1983).

66. G. E. Hinton and T. J. Sejnowski. Analyzing cooperative computation. In *Proceedings of the Fifth Annual Conference of the Cognitive Science Society, Rochester, NY, USA* (1983).

67. G. E. Hinton and T. J. Sejnowski. Optimal perceptual inference. In *Proceedings of the IEEE Conference on Computer Vision and Pattern Recognition, Washington, DC, USA*, vol. 448, pp. 448–453 (1983).

68. D. H. Ackley, G. E. Hinton, and T. J. Sejnowski, A learning algorithm for Boltzmann machines, *Cognitive Science.* **9**(1), 147–169 (1985).

69. P. Smolensky. Information processing in dynamical systems: Foundations of harmony theory. In *Parallel Distributed Processing: Explorations in the Microstructure of Cognition*, vol. 1, pp. 194–281. MIT Press, Cambridge, MA, USA (1986).

70. G. E. Hinton, Training products of experts by minimizing contrastive divergence, *Neural Computation.* **14**(8), 1771–1800 (2002).

71. G. E. Hinton, S. Osindero, and Y.-W. Teh, A fast learning algorithm for deep belief nets, *Neural computation.* **18**(7), 1527–1554 (2006).

72. G. E. Hinton and R. R. Salakhutdinov, Reducing the dimensionality of data with neural networks, *Science.* **313**(5786), 504–507 (2006).

73. D. Kim and D.-H. Kim, Smallest neural network to learn the Ising criticality, *Physical Review E.* **98**(2), 022138 (2018).

74. S. Efthymiou, M. J. Beach, and R. G. Melko, Super-resolving the ising model with convolutional neural networks, *Physical Review B.* **99**(7), 075113 (2019).

75. F. D'Angelo and L. Böttcher, Learning the Ising model with generative neural networks, *Physical Review Research.* **2**(2), 023266 (2020).

76. G. Carleo and M. Troyer, Solving the quantum many-body problem with artificial neural networks, *Science.* **355**(6325), 602–606 (2017).

77. G. Torlai, G. Mazzola, J. Carrasquilla, M. Troyer, R. Melko, and G. Carleo, Neural-network quantum state tomography, *Nature Physics.* **14**(5), 447–450 (2018).

78. J. Schmidhuber, Deep learning in neural networks: An overview, *Neural Networks.* **61**, 85–117 (2015).

79. J. Dean, A golden decade of deep learning: Computing systems & applications, *Daedalus.* **151**(2), 58–74 (2022).

80. S. Hochreiter and J. Schmidhuber, Long short-term memory, *Neural Computation.* **9**(8), 1735–1780 (1997).

81. Y. LeCun, L. Bottou, Y. Bengio, and P. Haffner, Gradient-based learning applied to document recognition, *Proceedings of the IEEE.* **86**(11), 2278–2324 (1998).

82. I. J. Goodfellow, J. Pouget-Abadie, M. Mirza, B. Xu, D. Warde-Farley, S. Ozair, A. C. Courville, and Y. Bengio. Generative adversarial nets. In eds. Z. Ghahramani, M. Welling, C. Cortes, N. D. Lawrence, and K. Q. Weinberger, *Advances in Neural Information Processing Systems 27: Annual Conference on Neural Information Processing Systems 2014, December 8-13 2014, Montreal, Quebec, Canada*, pp. 2672–2680 (2014).

83. A. Vaswani, N. Shazeer, N. Parmar, J. Uszkoreit, L. Jones, A. N. Gomez, L. Kaiser, and I. Polosukhin. Attention is all you need. In eds. I. Guyon, U. von Luxburg, S. Bengio, H. M. Wallach, R. Fergus, S. V. N. Vishwanathan, and R. Garnett, *Advances in Neural Information Processing Systems 30: Annual Conference on Neural Information Processing Systems 2017, December 4-9, 2017, Long Beach, CA, USA*, pp. 5998–6008 (2017).

84. A. Paszke, S. Gross, S. Chintala, G. Chanan, E. Yang, Z. DeVito, Z. Lin, A. Desmaison, L. Antiga, and A. Lerer, Automatic differentiation in PyTorch (2017).

85. G. Cybenko, Approximation by superpositions of a sigmoidal function, *Mathematics of Control, Signals and Systems.* **2**(4), 303–314 (1989).

86. K. Hornik, M. Stinchcombe, and H. White, Multilayer feedforward networks are universal approximators, *Neural Networks.* **2**(5), 359–366 (1989).

87. K. Hornik, Approximation capabilities of multilayer feedforward networks, *Neural Networks.* **4**(2), 251–257 (1991).

88. Y.-J. Wang and C.-T. Lin, Runge-Kutta neural network for identification of dynamical systems in high accuracy, *IEEE Transactions on Neural Networks.* **9**(2), 294–307 (1998).

89. T. Q. Chen, Y. Rubanova, J. Bettencourt, and D. Duvenaud. Neural ordinary differential equations. In eds. S. Bengio, H. M. Wallach, H. Larochelle, K. Grauman, N. Cesa-Bianchi, and R. Garnett, *Advances in Neural Information Processing Systems 31: Annual Conference on Neural Information Processing Systems 2018, NeurIPS 2018, December 3-8, 2018, Montréal, Canada*, pp. 6572–6583 (2018).

90. E. Kaiser, J. N. Kutz, and S. L. Brunton, Data-driven discovery of Koopman eigenfunctions for control, *Machine Learning: Science and Technology.* **2**(3), 035023 (2021).

91. F. Schäfer, P. Sekatski, M. Koppenhöfer, C. Bruder, and M. Kloc, Control of stochastic quantum dynamics by differentiable programming, *Machine Learning: Science and Technology.* **2**(3), 035004 (2021).

92. L. Böttcher, N. Antulov-Fantulin, and T. Asikis, AI Pontryagin or how artificial neural networks learn to control dynamical systems, *Nature Communications.* **13**(1), 333 (2022).

93. T. Asikis, L. Böttcher, and N. Antulov-Fantulin, Neural ordinary differential equation control of dynamics on graphs, *Physical Review Research.* **4**(1), 013221 (2022).

94. L. Böttcher and T. Asikis, Near-optimal control of dynamical systems with neural ordinary differential equations, *Machine Learning: Science and Technology.* **3**(4), 045004 (2022).

95. L. Böttcher, T. Asikis, and I. Fragkos, Control of dual-sourcing inventory systems using recurrent neural networks, *INFORMS Journal on Computing.* **35**(6), 1308–1328 (2023).

96. L. Böttcher. Gradient-free training of neural ODEs for system identification and control using ensemble Kalman inversion. In *ICML Workshop on New Frontiers in Learning, Control, and Dynamical Systems, Honolulu, HI, USA, 2023* (2023).

97. D. Ruiz-Balet and E. Zuazua, Neural ODE control for classification, approximation, and transport, *SIAM Review.* **65**(3), 735–773 (2023).

98. T. Asikis. Towards recommendations for value sensitive sustainable consumption. In *NeurIPS 2023 Workshop on Tackling Climate Change with Machine Learning: Blending New and Existing Knowledge Systems* (2023). URL `https://nips.cc/virtual/2023/76939`.

99. M. Raissi, P. Perdikaris, and G. E. Karniadakis, Physics-informed neural networks: A deep learning framework for solving forward and inverse problems involving nonlinear partial differential equations, *Journal of Computational physics.* **378**, 686–707 (2019).

100. S. L. Brunton and J. N. Kutz, *Data-driven science and engineering: Machine learning, dynamical systems, and control.* Cambridge University Press, Cambridge, UK (2022).

101. S. Mowlavi and S. Nabi, Optimal control of PDEs using physics-informed neural networks, *Journal of Computational Physics.* **473**, 111731 (2023).

102. C. Fronk and L. Petzold, Interpretable polynomial neural ordinary differential equations, *Chaos: An Interdisciplinary Journal of Nonlinear Science.* **33**(4) (2023).

103. M. Xia, L. Böttcher, and T. Chou, Spectrally adapted physics-informed neural networks for solving unbounded domain problems, *Machine Learning: Science and Technology.* **4**(2), 025024 (2023).

104. L. Böttcher, L. L. Fonseca, and R. C. Laubenbacher, Control of medical digital twins with artificial neural networks, *arXiv preprint arXiv:2403.13851* (2024).

105. L. Böttcher and G. Wheeler, Visualizing high-dimensional loss landscapes with Hessian directions, *Journal of Statistical Mechanics: Theory and Experiment.* p. 023401 (2024).

106. J. J. Ruiz-Lorenzo, *Nature of the Spin Glass Phase in Finite Dimensional (Ising) Spin Glasses,* In *Order, Disorder and Criticality, Volume 6* (ed. Yu. Holovatch), chapter 1, 1-52 (2020).

107. M. Thoma. LaTeX Examples: Hopfield Network. URL `https://github.com/MartinThoma/LaTeX-examples/tree/master/tikz/hopfield-network` (Accessed: September 16, 2023).

108. J. Hertz, R. G. Palmer, and A. S. Krogh, *Introduction to the Theory of Neural Computation*, 1st edn. Perseus Publishing, New York City, New York, USA (1991).

109. A. J. Storkey. Increasing the capacity of a Hopfield network without sacrificing functionality. In eds. W. Gerstner, A. Germond, M. Hasler, and J. Nicoud, *Artificial Neural Networks - ICANN '97, 7th International Conference, Lausanne, Switzerland, October 8-10, 1997, Proceedings*, vol. 1327, *Lecture Notes in Computer Science*, pp. 451–456, Springer (1997).

110. N. Metropolis, A. W. Rosenbluth, M. N. Rosenbluth, A. H. Teller, and E. Teller, Equation of state calculations by fast computing machines, *The Journal of Chemical Physics*. **21**(6), 1087–1092 (1953).

111. J. E. Gubernatis, Marshall Rosenbluth and the Metropolis algorithm, *Physics of Plasmas*. **12**(5) (2005).

112. S. Kirkpatrick, C. D. Gelatt Jr, and M. P. Vecchi, Optimization by simulated annealing, *Science*. **220**(4598), 671–680 (1983).

113. R. J. Glauber, Time-dependent statistics of the Ising model, *Journal of Mathematical Physics*. **4**(2), 294–307 (1963).

114. M. Á. Carreira-Perpiñán and G. E. Hinton. On contrastive divergence learning. In eds. R. G. Cowell and Z. Ghahramani, *Proceedings of the Tenth International Workshop on Artificial Intelligence and Statistics, AISTATS 2005, Bridgetown, Barbados, January 6-8, 2005*, Society for Artificial Intelligence and Statistics (2005).

115. G. Hinton, A practical guide to training restricted Boltzmann machines, *Momentum*. **9**(1), 926 (2010).

116. S. Park, C. Yun, J. Lee, and J. Shin. Minimum width for universal approximation. In *9th International Conference on Learning Representations, ICLR 2021, Virtual Event, Austria, May 3-7, 2021* (2021).

117. M. Hardt, B. Recht, and Y. Singer. Train faster, generalize better: Stability of stochastic gradient descent. In eds. M. Balcan and K. Q. Weinberger, *Proceedings of the 33nd International Conference on Machine Learning, ICML 2016, New York City, NY, USA, June 19-24, 2016*, vol. 48, *JMLR Workshop and Conference Proceedings*, pp. 1225–1234, JMLR.org (2016).

118. A. C. Wilson, R. Roelofs, M. Stern, N. Srebro, and B. Recht, The marginal value of adaptive gradient methods in machine learning. *arXiv preprint arXiv:1705.08292* (2017).

119. D. Choi, C. J. Shallue, Z. Nado, J. Lee, C. J. Maddison, and G. E. Dahl, On empirical comparisons of optimizers for deep learning, *arXiv preprint arXiv:1910.05446* (2019).

120. N. S. Keskar, D. Mudigere, J. Nocedal, M. Smelyanskiy, and P. T. P. Tang. On large-batch training for deep learning: Generalization gap and sharp minima. In *5th International Conference on Learning Representations, ICLR 2017, Toulon, France, April 24-26, 2017, Conference Track Proceedings* (2017).

121. L. Dinh, R. Pascanu, S. Bengio, and Y. Bengio. Sharp minima can generalize for deep nets. In eds. D. Precup and Y. W. Teh, *Proceedings of the 34th International Conference on Machine Learning, ICML 2017, Sydney,*

NSW, Australia, 6-11 August 2017, vol. 70, *Proceedings of Machine Learning Research*, pp. 1019–1028 (2017).

122. I. J. Goodfellow and O. Vinyals. Qualitatively characterizing neural network optimization problems. In eds. Y. Bengio and Y. LeCun, *3rd International Conference on Learning Representations, ICLR 2015, San Diego, CA, USA, May 7-9, 2015, Conference Track Proceedings* (2015).

123. H. Li, Z. Xu, G. Taylor, C. Studer, and T. Goldstein. Visualizing the loss landscape of neural nets. In eds. S. Bengio, H. M. Wallach, H. Larochelle, K. Grauman, N. Cesa-Bianchi, and R. Garnett, *Advances in Neural Information Processing Systems 31: Annual Conference on Neural Information Processing Systems 2018, NeurIPS 2018, December 3-8, 2018, Montréal, Canada*, pp. 6391–6401 (2018).

124. Y. Hao, L. Dong, F. Wei, and K. Xu. Visualizing and understanding the effectiveness of BERT. In eds. K. Inui, J. Jiang, V. Ng, and X. Wan, *Proceedings of the 2019 Conference on Empirical Methods in Natural Language Processing and the 9th International Joint Conference on Natural Language Processing, EMNLP-IJCNLP 2019, Hong Kong, China, November 3-7, 2019*, pp. 4141–4150, Association for Computational Linguistics (2019).

125. D. Wu, S. Xia, and Y. Wang. Adversarial weight perturbation helps robust generalization. In eds. H. Larochelle, M. Ranzato, R. Hadsell, M. Balcan, and H. Lin, *Advances in Neural Information Processing Systems 33: Annual Conference on Neural Information Processing Systems 2020, NeurIPS 2020, December 6-12, 2020, virtual* (2020).

126. S. Horoi, J. Huang, B. Rieck, G. Lajoie, G. Wolf, and S. Krishnaswamy. Exploring the geometry and topology of neural network loss landscapes. In eds. T. Bouadi, É. Fromont, and E. Hüllermeier, *Advances in Intelligent Data Analysis XX - 20th International Symposium on Intelligent Data Analysis, IDA 2022, Rennes, France, April 20-22, 2022, Proceedings*, vol. 13205, *Lecture Notes in Computer Science*, pp. 171–184, Springer (2022).

127. J. M. Lee, *Riemannian manifolds: An introduction to curvature.* vol. 176, Springer Science & Business Media, New York City, NY, USA (2006).

128. M. Berger and B. Gostiaux, *Differential Geometry: Manifolds, Curves, and Surfaces.* Springer Science & Business Media, New York City, NY, USA (2012).

129. W. Kühnel, *Differential Geometry.* American Mathematical Society, Providence, RI, USA (2015).

130. K. Crane, Discrete differential geometry: An applied introduction, *Notices of the AMS, Communication.* pp. 1153–1159 (2018).

131. J. Martens, New insights and perspectives on the natural gradient method, *Journal of Machine Learning Research.* **21**(146), 1–76 (2020).

132. R. Hecht-Nielsen, Context vectors: general purpose approximate meaning representations self-organized from raw data, *Computational Intelligence: Imitating Life, IEEE Press.* pp. 43–56 (1994).

133. R. B. Davies, Algorithm AS 155: The distribution of a linear combination of χ^2 random variables, *Applied Statistics.* **29**(3), 323–333 (1980).

134. B. Laurent and P. Massart, Adaptive estimation of a quadratic functional by model selection, *Annals of Statistics.* **28**(5), 1302–1338 (2000).

135. J. Bausch, On the efficient calculation of a linear combination of chi-square random variables with an application in counting string vacua, *Journal of Physics A: Mathematical and Theoretical.* **46**(50), 505202 (2013).

136. T. Chen and T. Lumley, Numerical evaluation of methods approximating the distribution of a large quadratic form in normal variables, *Computational Statistics & Data Analysis.* **139**, 75–81 (2019).

137. J. Matoušek, On variants of the Johnson–Lindenstrauss lemma, *Random Structures and Algorithms.* **33**(2), 142–156 (2008).

138. M. F. Hutchinson, A stochastic estimator of the trace of the influence matrix for Laplacian smoothing splines, *Communications in Statistics-Simulation and Computation.* **18**(3), 1059–1076 (1989).

139. Z. Bai, G. Fahey, and G. Golub, Some large-scale matrix computation problems, *Journal of Computational and Applied Mathematics.* **74**(1-2), 71–89 (1996).

140. H. Avron and S. Toledo, Randomized algorithms for estimating the trace of an implicit symmetric positive semi-definite matrix, *Journal of the ACM (JACM).* **58**(2), 1–34 (2011).

141. Y. N. Dauphin, R. Pascanu, Ç. Gülçehre, K. Cho, S. Ganguli, and Y. Bengio. Identifying and attacking the saddle point problem in high-dimensional non-convex optimization. In eds. Z. Ghahramani, M. Welling, C. Cortes, N. D. Lawrence, and K. Q. Weinberger, *Advances in Neural Information Processing Systems 27: Annual Conference on Neural Information Processing Systems 2014, December 8-13 2014, Montreal, Quebec, Canada,* pp. 2933–2941 (2014).

142. P. Baldi and K. Hornik, Neural networks and principal component analysis: Learning from examples without local minima, *Neural Networks.* **2**(1), 53–58 (1989).

143. Z. Yao, A. Gholami, K. Keutzer, and M. W. Mahoney. PyHessian: Neural networks through the lens of the Hessian. In eds. X. Wu, C. Jermaine, L. Xiong, X. Hu, O. Kotevska, S. Lu, W. Xu, S. Aluru, C. Zhai, E. Al-Masri, Z. Chen, and J. Saltz, *2020 IEEE International Conference on Big Data (IEEE BigData 2020), Atlanta, GA, USA, December 10-13, 2020,* pp. 581–590, IEEE (2020).

144. Z. Liao and M. W. Mahoney. Hessian eigenspectra of more realistic nonlinear models. In eds. M. Ranzato, A. Beygelzimer, Y. N. Dauphin, P. Liang, and J. W. Vaughan, *Advances in Neural Information Processing Systems 34: Annual Conference on Neural Information Processing Systems 2021, NeurIPS 2021, December 6-14, 2021, virtual,* pp. 20104–20117 (2021).

145. R. L. Burden, J. D. Faires, and A. M. Burden, *Numerical Analysis; 10th edition.* Cengage Learning, Boston, MA, USA (2015).

146. B. Adcock and N. Dexter, The gap between theory and practice in function approximation with deep neural networks, *SIAM Journal on Mathematics of Data Science.* **3**(2), 624–655 (2021).

https://doi.org/10.1142/9789819800827_0004

Chapter 4

Political Systems as Complex, Statistical Systems – Unravelling the Role of Diversity and Feedback for Stability

Karoline Wiesner[*]

*Institute of Physics and Astronomy,
University of Potsdam, 14469 Potsdam, Germany*

Social systems are complex systems, and, although not physical systems in the traditional sense, they can be discussed surprisingly clearly with the concepts of statistical physics. We review the perspective of social, political systems as complex systems and summarise the role of diversity and feedback in their stability. A recent quantitative study of data provided by the Varieties of Democracy project is reviewed. The results are discussed in light of the role of diversity and feedback for stability of complex social systems.

Contents

1. Introduction 164
2. Stability of democracy 165
3. Features of complex systems 167
   3.1. Stability in complex systems 168
   3.2. Stability in statistical mechanics 169
   3.3. Stability vs disorder and feedback 170
4. The Varieties of Democracy data 170
   4.1. Principal component analysis of the V-Dem data 171
   4.2. Potential feedback mechanism stabilising autocracies and democracies . 173
5. Stability of democracy from a complex-systems perspective 174
   5.1. Disorder and randomness 174
   5.2. Diversity 176
   5.3. Feedback 177
6. Conclusion 177
References 178

---

[*]karoline.wiesner@uni-potsdam.de, ORCID: https://orcid.org/0000-0003-2944-1988

1. Introduction

In societal dynamics, where an environment of politics and economics interweave with the complexities of human behaviour, lies a field of quantitative political-science research through the lens of complexity science. It is imperative to understand the undercurrents shaping stability, robustness, and resilience within governance and democracy. Drawing upon principles from statistical physics and complexity science, we have begun to glean valuable insights into the functioning of democracies and the mechanisms that underpin their stability.[1]

At the heart of this exploration lies the concept of stability—a notion deeply ingrained in both physics, complexity, and political science. The difficulty of assessing stability of political systems is illustrated by the decision of the Axel Springer company, a prominent publisher in the German media landscape in the last decades of the 20th century. They insisted, during the 70's and 80's, keeping quote signs around the acronym for the German Democratic Republic ('GDR', rather than GDR) to communicate his conviction that reunification of the two Germanies must and will happen. In August 1989 they decided to give up and to lift the quotes.[2,3] We all know what happened only two months later: On 9 November that same year the Berlin wall came down, in a peaceful way. And only one year later, the two Germanies were reunited.

In this contribution, we will bring together several strands of investigation on the stability of democracy from the perspective of complexity science. The emphasis of the following discussion is on three elements, or rather three features, of complex systems as defined by Ladyman and Wiesner[4]—disorder, feedback, and robustness—and their relevance to stability of a social system, such as that of a democratic society. We will review existing studies and add to those a discussion on statistical physics as a source of relevant concepts and tools for studying the stability of social and political systems.

There are major concerns about the possibility of the world becoming less democratic that we do not fully understand. Available data that provide estimates of 'democraticness' show that democracy scores have been in decline for over a decade. In Section 2, we will give a brief overview over these recent developments, and over some of the existing literature. We will review the features of complex systems and the ways in which a democracy is a complex system in Section 3. The central feature of robustness will be highlighted, including its meaning in the context of democracy.

A democracy is constituted of living beings, and, yet, concepts of statistical physics, in particular non-equilibrium statistical physics, will be helpful to approach the stability of such systems. We will give an outlook on the potential research questions that could be asked taking on a statistical physics perspective. On the back of these reviews, we will summarise and discuss results of a statistical analysis of data provided by the political-science project 'Varieties of Democracy' (Section 4).

Thanks to the increasing availability of quantitative data, we are able to take a first step in the direction of actually linking phenomena known from statistical physics, such as temperature driven phenomena, feedback, and extreme events, to the dynamics in political systems.

2. Stability of democracy

The foundational idea of a democracy is that it provides a public market place of ideas, however imperfect, where competing positions are discussed and decided upon.

By the end of the 20th century, democracy was globally on the rise and there was broad agreement that democracy was the political ideal to strive for. Democratic countries were, on average, richer, less likely to enter war, and more apt at fighting corruption. However, in 2005 global democracy levels began to stagnate, and since 2015 they have been in decline (see Fig. 1 for quantitative evidence provided by the V-Dem data). Why democracy

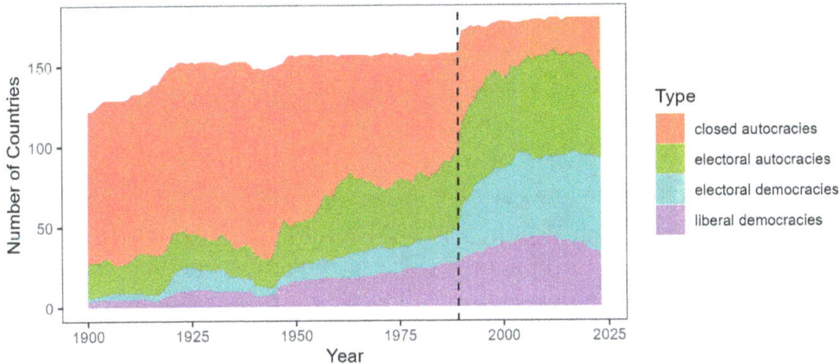

Figure 1. Political regimes based on the classification by Lührmann *et al.*,[5] showing V-Dem data aggregated by Our World in Data.[6] The dashed line indicates the year 1989.

is losing momentum, and whether this is only temporary, is almost entirely unknown. Standard approaches to addressing this question have so far failed to reach any clear conclusions. The perception of democracy being under threat has surged significantly in recent years. The Economist magazine calls the year 2024 "a giant test of nerves for democracy."[7]

The Economist identified 80 countries, including the USA and several consolidated European democracies, whose democracy scores declined over the last decade.[7] While much research has focused on how democracies arise and can be facilitated, there is limited understanding of how democracies destabilize. The conventional assumption has been that achieving democracy is a one-way ratchet, but only recently has the issue of 'democratic backsliding' garnered research attention.[8]

While the last two centuries have seen numerous 'democratic transitions'—non-democratic countries becoming democratic—instances of democracies transitioning to non-democratic regimes have been less frequent and less studied. Political scientists Waldner and Lust highlight this gap, noting a lack of theories to explain democratic backsliding.[8] The circumstances leading to democratic instability are under-researched, often only explored through single-case studies without a robust theoretical framework. This reflects a historical complacency within the West, where the stability of democratic institutions has been largely taken for granted over the last 50–60 years.

Recent events in Europe, such as in Poland and Hungary, where elected officials have undermined critical media and key institutions like independent courts, challenge the assumption that democracies are inherently stable. The political sciences have termed this phenomenon 'democratic backsliding'. Identifying instances of democratic backsliding is inherently controversial and politically charged.

Opposite to the slow change implied by the term 'backsliding' are fast changes, such as the collapse of the Soviet Union. Political scientist Yascha Mounk highlighted the tendency among scholars to overlook the possibility of extreme events, underscoring the need for a nuanced understanding of stability within political systems.[3] The introductory example is proving the opposite. Extreme events are different from backsliding, of course. However, what they have in common is that for either of these political scientists lack theories to explain. By summarising existing recent studies on these questions, we hope to provide a roadmap and motivation to further investigate the usefulness of complexity science and statistical mechanics in the search for theories of political instability.

We begin our journey by reviewing features of complex systems, which will lead us on to think about the origins of stability in complex systems. From here we will make the Segway to statistical mechanics and its notion of stability.

3. Features of complex systems

The theory of complex systems offers a collection of concepts and methods that bridge disciplines and highlight processes acting across time and length scales. Popular accounts of complex systems can be found in works by Holland,[9] Gell-Mann,[10] and Mitchell,[11] while more rigorous treatments are provided by Strevens,[12] Ladyman and Wiesner,[4] and Holovatch *et al.*[13] A minimal definition of a complex system is a collection of many elements with repeated interactions, exhibiting self-organization—sustaining patterns without central or external control.[4] In living systems, additional features such as hierarchical organization and adaptive behavior are observed. In the following, we list the features exhibited by complex systems, as identified by Ladyman and Wiesner[4] and give some detail on each.

Large number of parts and interactions: The number of parts, or elements, that constitute the system and the large number of interactions between them are the basic components of any complex system. Diversity and disorder: Complex systems are heterogeneous in their types of elements and interactions. For example, the the up-and-coming science of urban systems studies the many components of a city including the inhabitants, the businesses, cars, streets, trees, governing institutions, and all the interactions that take place between these elements. Openness: Complex systems are never completely isolated from their environment but are open to external influences. Examples are the economy and climate influencing a society. Drawing the boundary between the system and its environment can be difficult and is to some extent a choice made by the observer. Feedback: Feedback is the phenomenon of past interactions influencing future interactions. Feedback loops can be stabilizing (negative) or self-reinforcing (destabilizing, positive). An example of stabilizing feedback is the temperature regulation of the lake's water surface through convection and evaporation. Feedback in a system can make it challenging to identify causalities and often means that causality is no longer a linear chain. Non-linearity: This is the non-linear dependence of one variable on another. A well-known example of nonlinearity is the presence of power-law frequency distributions in statistics of social systems; one of the earliest examples is the distribution of sizes of cities which over many orders of magnitude follows a power-law.[14] Another example of non-linearity are sudden transitions, so-called tipping points, for example in ecosystems.[15] A statistical correlation between two variables is also sometimes referred to as nonlinearity. Self-organization: This is the process by which spatial or dynamic structure is formed by a system through the many interactions between its parts, without a central controller. The spatial structure of the Grand Canyon, for example, is the result of the many interactions between water, rock, soil and microorganisms. Robustness and resilience: These terms are often used interchangeably, but they have a somewhat different meaning. Robustness is, technically, the insensitivity to changing conditions and is usually used in the context of modelling assumptions or input to algorithms. In the context of environmental modelling, the robustness of an intervention is discussed with respect to variations in assumptions being made about the system. On the other hand, resilience is a system's ability to withstand or readjust after a perturbation to preserve a function, such as life. Social systems can exhibit resilience against species loss through adaptive behaviour of individual species. Adaptive behaviour: This is the purposeful change of behaviour of a system in response to changing circumstances while the system maintains a particular function. Memory: This is the ability of a system to remember its past and use it for future action beneficially. Memory can exist in a group as much as in individuals. Modularity: This is the division of a system into groups or modules performing separate functions. An example is the structure of social groups. Smaller clusters of friends and family form one larger cluster of a local community. Many local communities form the bigger structure of a cities or regions. Adaptive behaviour: Examples of social complex systems with adaptive behavior include eusocial societies of bees, termites, ants, and wasps, extensively studied for their behavioral dynamics. Below we will concentrate on those complex-system features that are central for our analysis.

3.1. *Stability in complex systems*

Traditionally, stability has been viewed through the dichotomy of static and dynamic stability. In the former, a system remains insensitive to perturbations, while in the latter, the system returns to its equilibrium state following disturbances. The dynamics of complex systems introduce a layer of intricacy, where stability emerges from disorder and feedback loops. In such systems, interactions are decentralized, and diversity among components fosters resilience against disruptions. As such, stability in complex systems transcends mere resistance to change; it arises dynamically from the interplay of disorder and feedback.[4] Disorder, inherent in the decentralized nature of complex systems, fosters adaptability and innovation. In a democratic context, disorder manifests as the diversity of opinions, ideologies, and interests, which fuels healthy debate and prevents the stagnation of ideas.[1] Feedback loops, iterated interactions that influence subsequent behaviors, are instrumental in shaping emergent dynamics within complex systems. In democracies, feedback mechanisms such as elections, public discourse, and institutional checks and balances facilitate course corrections and ensure the responsiveness of governance to societal needs.

3.2. *Stability in statistical mechanics*

Complex systems and statistical physics system have in common that they consists of many individual elements, each of them having their individual dynamics while potentially interacting with each other. In both cases, the environment plays a crucial role, since it provides energy, matter or information to the system. Individuals in the social system might be subject to economic globalisation to government repression, for example, and they react to the environment, trying to find an equilibrium state.

In statistical mechanics, stability is linked to equilibrium states where the system spends most of its time in the long run. This equilibrium is robust, as it encompasses the largest number of microscopic configurations, and stable, as the system naturally returns to this state after perturbations.

This equilibrium concept aligns with thermodynamic principles where state functions are invariant. However, applying this to social systems is challenging, as social systems are not typically in thermal equilibrium and their dynamics often defy long-term predictions.

Mathematical models, including those from statistical mechanics, are often used to predict such regime changes. For instance, Sinha and Pan[16] employed the Ising model of ferromagnetism to explain sudden popularity shifts in ideas or products, mirroring real-world phenomena like election outcomes and movie popularity.

Drawing parallels with the Ising model of magnetization in physics, where phase transitions are driven by temperature and interaction, we can discern how critical transitions in democratic systems may be influenced by a confluence of factors. From a dynamical systems perspective, the mathematics of tipping points elucidates how seemingly minor perturbations can lead to drastic shifts in system behavior, providing, at the least, an analogy for sudden transformations observed in political landscapes.

We believe that the concepts of non-equilibrium statistical mechanics open an opportunity to understand the dynamics of political systems – in particular the observation of extreme events such as that of the dissolution of the GDR.

3.3. *Stability vs disorder and feedback*

In our discussion, we focus on three features of complex social systems, particularly democratic systems: disorder, feedback and stability, and their mutual relationship. Negative feedback stabilizes a system, such as a thermostat regulating room temperature or supply and demand dynamics in economics reaching equilibrium prices. Positive feedback, conversely, is destabilizing—enhancing rather than desirable. Examples include the accelerating melting of polar ice caps due to reduced albedo effect and the recruiting process in honey bee societies where scout bees exponentially attract more scouts until the entire hive moves to a new site. Financial bubbles also demonstrate positive feedback mechanisms.[17]

Complex systems science, now a mature field within the natural sciences, offers tools to address significant questions about social systems' robustness, including democracy. This perspective aims to draw attention to how complex systems theory can provide insights into political science's timely and crucial questions.

4. The Varieties of Democracy data

To support the qualitative discussion so far, we now review some results of a quantitative study on democracy and autocracy by Wiesner, Bien, and Wilson.[18]

In Ref. [18], Wiesner, Bien, and Wilson utilized the publicly available dataset on electoral democracy provided by the Varieties of Democracy (V-Dem) project. V-Dem is a large-scale collaborative initiative that employs expert coding to create quantitative data on various democratic attributes for over 190 countries, with historical data spanning decades to centuries, depending on the country, and annual updates. This project surveys numerous country experts and uses a Bayesian measurement model to estimate latent values. The surveys typically ask experts to rate the strength or openness of institutions, such as election intimidation or media censorship, on an ordinal scale (*e.g.*, "low," "intermediate," "high"). The measurement model then assesses the reliability between respondents and generates point estimates for each question.[19]

The V-Dem project aggregates information on different attributes into mid-level indices representing specific concepts like civil liberties and election quality, which collectively represent democracy. One primary index published by V-Dem is the Electoral Democracy Index (EDI). The EDI is based on quantitative data from over forty variables related to electoral democracy. The authors used relevant data from 1900 to 2021, totaling 12,296 data points (country-year events). Further details on the EDI's construction and the expert questions for each of the 24 variables are provided in Ref. [18] and references therein.

4.1. *Principal component analysis of the V-Dem data*

To evaluate the extent to which there are underlying dimensions in the data, Wiesner, Bien, and Wilson performed a principal component analysis (PCA).[18] PCA is one of the most simple and robust techniques for dimensionality reduction. It is part of a family of statistical techniques for representing high-dimensional data on a lower-dimensional linear subspace with as little information loss as possible. The data were normalized (*i.e.* centered to a mean of zero and rescaled to a variance of one). The principal component values were then rescaled to values within [-1,1] while preserving the mean of zero for all components.

The data projected onto the first two principal components are shown in Fig. 2 (reproduced from Ref. [18]). Examples of country-year data points are shown in the lower part of Fig. 2. The notable result of this PCA was the discovery of a strong first principal component (PC1) explaining 68% of the variance. An additional 10% is contained in the second principal component (PC2). Furthermore, the first principal component correlates strongly with the Electoral Democracy Index (EDI), which is constructed by hand by the V-Dem Team. Hence, one of the conclusions of the authors is that the EDI is a well constructed, informative index.

In the V-Dem dataset, high scores indicate a greater level of democracy, while low scores indicate less democracy. The association between the first two principal components (PC1 and PC2) and the V-Dem variables revealed that PC1 primarily represents democratic qualities tied to electoral competition and respect for civil rights and liberties. In contrast, PC2 highlights the balance between a state's ability to conduct effective

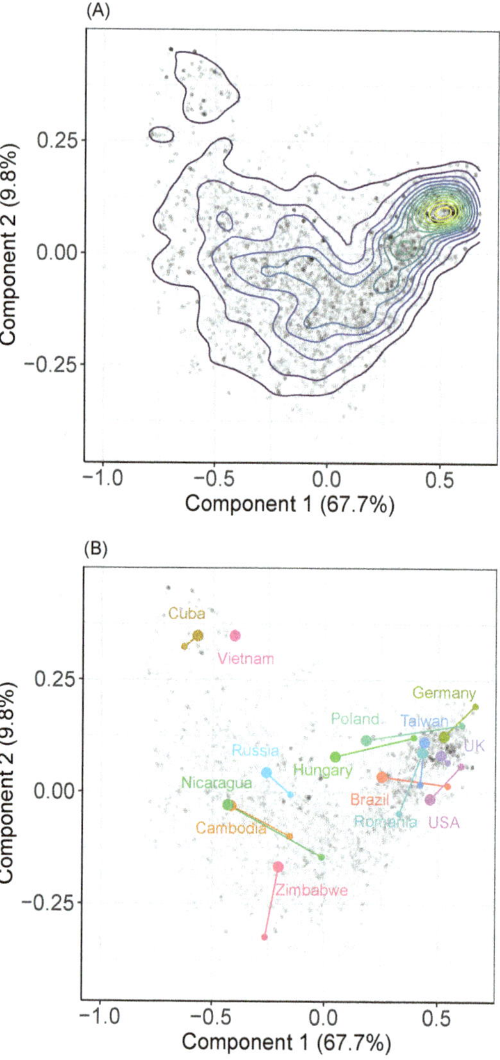

Figure 2.   All 12,296 country-year data points are plotted according to their PC1 and PC2 values. The upper panel (A) shows a kernel-density estimate, indicating the highest concentration of observations at relatively high PC1 (0.5) and moderately high PC2 (2.4), suggesting a strong democratic presence with effective electoral control. However, there is considerable variability, with some data points scoring very low on PC1 and very high on PC2.   The lower panel (B) plots example countries, showing electoral autocracies like Cuba and Vietnam in the upper-left area with tight electoral control but weak civil liberties. Some democratic regimes have shown backsliding, moving away from the populous area between 2011 and 2021. Figure reproduced from Ref. [18].

elections and its respect for civil liberties.

Since theses results indicated a nonlinearity in the variables composing the V-Dem democracy index, in a next step, Wiesner, Bien, an Wilson separated out the variables that loaded positive onto PC2 versus those that loaded negatively. Thus, they discovered a turning point on the way to democracy, which we review in the following.

4.2. *Potential feedback mechanism stabilising autocracies and democracies*

In the PCA, the second principal component turned out to have a clear pattern of all 'election-quality' variables loading positively onto it, while all variables measuring civic freedoms (related to 'free association' and 'free expression' plus multi-party elections) loaded negatively onto it.[18] When separating the V-Dem variables according to their weight onto PC2, those loading negatively on PC2 and those loading positively, a clear pattern emerged:

- For low PC1 values, variables related to 'election quality' fluctuate around a constant mean until a turning point around PC1 $\approx$ 0, above which their average increases.
- Variables related to 'free association' and 'free expression' continuously increase with higher PC1 values, showing narrowly distributed values.

The turning point in PC2 around PC1 $\approx$ 0 marks where election quality begins to rise above a minimal level required for (semi-)democratic regimes. Countries scoring low on PC1 include non-democracies with co-ordinated elections but restricted civil liberties, while higher PC1 scores tend to encompass both high election quality and civil liberties. As the authors point out, the dip in PC2 from low to high PC1 values suggests a trade-off between election control and civil liberties, indicating that as countries develop democratically, the interaction between election quality and civil liberties shifts from mutually suppressing to mutually enhancing.

Contrary to the assumption that election quality and civil liberties generally improve together, there is a threshold at which their correlation flips from negative to positive. This indicates a potential trade-off between the order provided by controlled elections and the liberty associated with democracy, explaining why some areas of the two-dimensional space are unoccupied.

This shift in trend clearly visible in the principal component data might be an indication of positive vs negative feedback. In the high-democracy region of the data (high values of PC1), clearly both electoral variables and variables of civic freedom and civic association are increasing together. This is not a proof of a mutual enhancement, but it is an indication that there is the potential for positive feedback between the two. At the other end of the democracy spectrum, in regions of low PC1 values, the situation appears to be the opposite. Here the electoral variables seem to rise when levels of civic freedoms and civil liberties decrease. Fig. 3 only shows the averages plus the variance. But it is clearly visible that there is a shift in mean and variance around the midpoint of the PC1 axis. Hence, there is an interesting separation between a regime in which both indicators rise together on a regime in which indicators are anti-correlated. This suggests that there is one regime with a positive feedback loop, strengthening democracy, and one regime with a suppressing mechanism of civic freedoms. Note, that there is no time information in the principal component analysis. Hence, there is the potential for deeper insights, once time is taken into account.

5. Stability of democracy from a complex-systems perspective

In a highly interdisciplinary collaboration, Wiesner *et al.* investigated the use of complexity science to understand stability of democracy.[1] Given the above discussion on the potential mechanisms for stabilising democracies and autocracies, we will now review these arguments to further establish and strengthen the link between concept of statistic of physics complexity science and stability in democracy.

5.1. *Disorder and randomness*

Random interactions are crucial for self-organization. For example, human group performance can improve with the inclusion of a few randomly behaving agents.[20] Democracy thrives on the unstructured exchange of opinions, as randomness prevents stagnation in bad ideas. David Runciman emphasizes that democracy's inherent randomness ensures continuous disruption and renewal.[21]

The role of this order is implicit in the above reviewed analysis of the V-Dem data. It is not unreasonable to say that in the presence of tightly controlled elections, with little uncertainty on the outcome, the amount of disorder in the system is very limited. Disorder here is understood in the

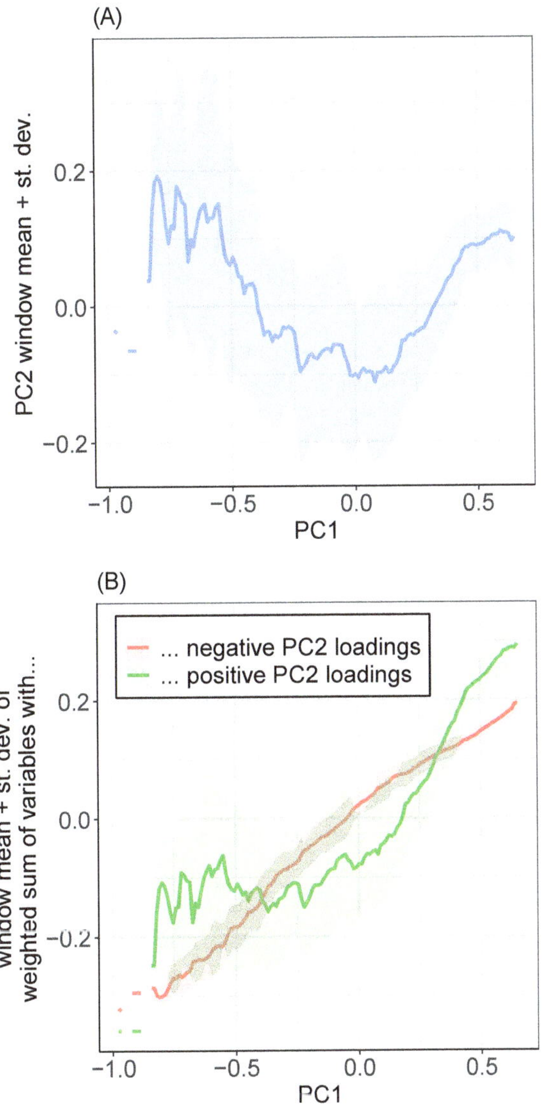

Figure 3. (A) The mean score of PC2 in a sliding window of PC1 with a width of 0.4 and a step size of 0.1. (B) The mean score of a decomposition of PC2 –
Green curve: (the weighted sum of) variables relating to 'election capability' and suffrage, all of which positively correlate with PC2. Red curve: (the weighted sum of) variables relating to civil liberties ('free association', 'free expression'), all of which negatively correlate with PC2. Thus, the green curve minus the red curve equals the blue curve in (A). Figure reproduced from Ref. [18].

sense discussed above: as diversity of opinions, diversity of parties that can be voted for, diversity of life styles that are allowed *etc.* The connection to statistical mechanics, just to continue a little with our speculation, is found here. Disorder in the statistical mechanics sense is, of course, temperature. A rise in temperature leads to a rise in the number of accessible micro states and also generally in the number of visited micro states within a given time window. If the diversity of paths towards a solution to any given problem is limited, then a system is stuck in a given state and has little ability to adapt to changing circumstances. The increase in temperature can also however mean the accessibility of stable states this of course is the idea underline the technique of simulated annealing for finding equilibria or optima of any given utility function. One might think of regimes with tightly controlled elections, with little uncertainty on the outcome, as 'quenched states', *i.e.* states that are frozen into a single configuration and that offer no mechanism to adapt to changing external conditions.

The power of randomness, a hallmark of complex systems, plays a pivotal role in shaping the dynamics of democracies. Shalizi and Farrell studied the influence of randomness in a hierarchical model of decision making.[22] Their model illustrate how an element of randomness can give access to more optimal solutions. This is, in spirit, very similar to the well-known numerical technique of simulated annealing.

As political scientist David Runciman aptly observes, the randomness inherent in democracy safeguards against the entrenchment of detrimental ideologies.[21] In the sam spirit, Farrell and Shalizi underscore the role of diversity and partisan disagreement in fostering healthy debate and innovation.

Ancient Athenian democracy offers an interesting historical example for the role of randomness and diversity in governance. The system randomly selected amongst its citizens (Greek adult male only) for public service and magistracy. Athenian democracy endured for over three centuries, to which the participation model of random selection might well have contributed.

5.2. *Diversity*

Diversity enhances the stability of complex systems, as seen in genetically diverse bee hives. Similarly, democracy benefits from diverse viewpoints, though too much or too little diversity can lead to instability. Echo chambers and selective information sharing on social media can create antagonistic sub-groups, undermining democratic agreement. The spread of misinformation via the internet exacerbates this issue, leading to political apathy or convergence on simplified propaganda.

5.3. *Feedback*

Complexity science, which is intricately linked to statistical mechanics, informs our understanding of stability through concepts like basins of attraction. Democratic institutions are seen as self-enforcing equilibria, fostering dynamic activities without being destabilized themselves. Game theory, applied to social norms, helps analyze the strategic propagation and erosion of norms, offering insights into the resilience of democratic institutions.

Feedback mechanisms are crucial in complex systems, leading to phenomena like wealth inequality translating into political power. This feedback can transform democracy into oligarchy if unchecked. However, high inequality can also spur counter-mobilization for political change, provided citizens have sufficient knowledge and access to public discourse.[23]

In Ref. [24], Eliassi-Rad *et al.* explore policy implications of complexity science for democratic governance. Several lessons in how complexity theory can inform decision-making processes were given, with the aim of enhancing stability and quality of democratic governance.

Complexity science, by treating social systems as dynamical systems, helps us understand the decay of pro-democratic patterns and the rise of alternative self-reinforcing phenomena. For example, in opinion dynamics, consistent and competent minorities can significantly influence outcomes, supporting Tocqueville's view that "More fires get started in a democracy but more fires get put out, too".[25]

6. Conclusion

In conclusion, the physics of complex systems provides a rich framework for studying the intricacies of human behaviour in the context of democracy, and of politics in general. By unravelling the interplay between stability, randomness, and diversity, we can glean valuable lessons for supporting resilient democratic systems. As we face new and old crises in an increasingly interconnected world, we should remember that science can identify ways forward in our strive to build free and stable societies that thrive amidst uncertainty and change.[24]

References

1. K. Wiesner, A. Birdi, T. Eliassi-Rad, H. Farrell, D. García, S. Lewandowsky, P. Palacios, D. Ross, D. Sornette, and K. Thébault, Stability of democracies: a complex systems perspective, *European Journal of Physics.* **40**(1), 014002 (2018).
2. Los Angeles Times, Bonn publisher lifts quotes on DDR, *Los Angeles Times.* **2 August** (1989).
3. Y. Mounk, *The Great Experiment: Why Diverse Democracies Fall Apart and How They Can Endure.* Penguin (2022).
4. J. Ladyman and K. Wiesner, *What is a complex system?* Yale University Press (2020).
5. A. Lührmann, S. I. Lindberg, and M. Tannenberg, Regimes in the world (riw): A robust regime type measure based on v-dem, *V-Dem Working Paper.* **47** (2017).
6. B. Herre, L. Rodés-Guirao, E. Ortiz-Ospina, and M. Roser, Democracy, *Our World in Data* (2013). https://ourworldindata.org/democracy.
7. The Economist, 2024 is a giant test of nerves for democracy, *The Economist.* **11 Feburary** (2024). `https://www.economist.com/international/2024/02/11/2024-is-a-giant-test-of-nerves-for-democracy`.
8. D. Waldner and E. Lust, Unwelcome change: Coming to terms with democratic backsliding, *Annual Review of Political Science.* **21**, 93–113 (2018).
9. J. H. Holland, Complex adaptive systems, *Daedalus.* **121**(1), 17–30 (1992).
10. M. Gell-Mann, *The Quark and the Jaguar: Adventures in the Simple and the Complex.* Macmillan (1995).
11. M. Mitchell, *Complexity: A Guided Tour.* Oxford Univerity Press (2009).
12. M. R. Strevens, *Bigger than chaos: The probabilistic structure of complex systems.* Rutgers, New Jersey (1996).
13. Y. Holovatch, R. Kenna, and S. Thurner, Complex systems: physics beyond physics, *European Journal of Physics.* **38**(2), 023002 (2017).
14. F. Auerbach, Das Gesetz der Bevölkerungskonzentration [The law of population concentration], *Petermanns Geographische Mitteilungen.* **59**, 74–76 (1913).
15. M. Scheffer, S. R. Carpenter, T. M. Lenton, J. Bascompte, W. Brock, V. Dakos, J. Van de Koppel, I. A. Van de Leemput, S. A. Levin, E. H. Van Nes, et al., Anticipating critical transitions, *Science.* **338**(6105), 344–348 (2012).
16. S. Sinha and R. K. Pan, How a hit is born: The emergence of popularity from the dynamics of collective choice, *Econophysics and Sociophysics: Trends and Perspectives.* **2**, 417–447 (2006).
17. D. Sornette and P. Cauwels, Financial bubbles: mechanisms and diagnostics, *arXiv preprint arXiv:1404.2140* (2014).
18. K. Wiesner, S. Bien, and M. C. Wilson, The principal components of electoral regimes–separating autocracies from pseudo-democracies, *arXiv preprint arXiv:2402.11335* (2024), to appear in Royal Society Open Science.
19. M. Coppedge, J. Gerring, C. H. Knutsen, S. I. Lindberg, J. Teorell, D. Alt-

man, M. Bernhard, A. Cornell, M. S. Fish, L. Gastaldi, H. Gjerløw, A. Glynn, S. Grahn, A. Hicken, K. Kinzelbach, K. L. Marquardt, K. McMann, V. Mechkova, P. Paxton, D. Pemstein, J. v. Römer, B. Seim, R. Sigman, S.-E. Skaaning, J. Staton, E. Tzelgov, L. Uberti, Y.-t. Wang, T. Wig, and D. Ziblatt. V-dem codebook v12. Technical report, Varieties of Democracy (V-Dem) Project (2022).

20. H. Shirado and N. A. Christakis, Locally noisy autonomous agents improve global human coordination in network experiments, *Nature.* **545**(7654), 370–374 (2017).

21. Runciman, David, Why replacing politicians with experts is a reckless idea, *The Guardian.* **1 May** (2018). `https://www.theguardian.com/news/2018/may/01/why-replacing-politicians-with-experts-is-a-reckless-idea`.

22. H. Farrell and C. R. Shalizi, *From voice to influence: Understanding citizenship in a digital age*, chapter Pursuing cognitive democracy. The University of Chicago Press Chicago, IL, USA (2015).

23. B. I. Page and M. Gilens, *Democracy in America?: What has gone wrong and what we can do about it.* University of Chicago Press (2020).

24. T. Eliassi-Rad, H. Farrell, D. Garcia, S. Lewandowsky, P. Palacios, D. Ross, D. Sornette, K. Thébault, and K. Wiesner, What science can do for democracy: a complexity science approach, *Humanities and Social Sciences Communications.* **7**(1), 1–4 (2020).

25. Alexis de Tocqueville as cited in Ref. [21].

https://doi.org/10.1142/9789819800827_0005

Chapter 5

Observing Cities as a Complex System

Rafael Prieto-Curiel*

Complexity Science Hub, Josefstädter Str. 39, A-1080, Vienna, Austria

Cities are some of the most intricate and advanced creations of humanity. Most objects in cities are perfectly synchronised to coordinate activities like jobs, education, transportation, entertainment or waste management. Although each city has its characteristics, some commonalities can be observed across most cities, like issues related to noise, pollution, segregation and others. Further, some of these issues might be accentuated in larger or smaller cities. For example, with more people, a city might experience more competition for space so rents would be higher. The urban scaling theory gives a framework to analyse cities in the context of their size. New data for analysing urban scaling theory allow an understanding of how urban metrics change with their population size, whether they apply across most regions or if patterns correspond only to some country or region. Yet, reducing a city and all its complexity into a single indicator might simplify urban areas to an extent where their disparities and variations are overlooked. Often, the differences between the living conditions in different parts of the same city are bigger than the degree of variation observed between cities. For example, in terms of rent or crime, within-city variations might be more significant than between cities. Here, we review some urban scaling principles and explore ways to analyse variations within the same city.

Contents

1. Introduction 182
2. How to analyse cities using complex systems 183
3. Urban scaling 187
   3.1. Quantifying urban scaling 189
   3.2. Why do we observe urban scaling? 190
4. Scaling infrastructure of cities 191
   4.1. The Line in Saudi Arabia 195
   4.2. Urban scaling and ownership 196
   4.3. The challenges of urban scaling studies 200

*prieto-curiel@csh.ac.at, ORCID: https://orcid.org/0000-0002-0738-2633

5. Urban studies beyond scaling 202
    5.1. Comparing at a smaller scale between cities 203
6. Conclusions 206
References 207

1. Introduction

The world's population has undergone rapid growth in recent decades, and this trend will not change for decades.[1] Some countries, such as Niger or Chad, will double their 2020 population before 2050.[2] This population growth mostly happens in cities. In fact, the rural population worldwide has already reached its peak size (of nearly 3.4 billion people) and will decrease in the upcoming decades.[3] Thus, although the world's population will keep increasing, that process will only occur in cities. Whilst many parts of the world have already undergone urbanisation, the next three decades will bring sweeping changes in many parts, including Africa and South Asia.[4–6] Due to the urbanisation process and population growth, some cities will keep growing at an unprecedented speed and might reach a population of 80 million inhabitants or more.[7] How could a city with 80 million inhabitants function? What shape will it have, or what surface will it occupy? How long will people spend on their daily commute?

Cities offer unparalleled access to a wide array of essential services, including healthcare, education, transportation, cultural amenities, water and sanitation services.[8–11] For example, in 2022, 81% of the world's urban population had access to safely managed drinking water services, but only 62% of the rural population had access to it.[12] Similarly, 65% of the world's urban population had access to safely managed sanitation services, against 46% of the world's rural population. The difference between urban and rural populations is even more marked in poorer parts of the world. Cities are the hubs where the integration of services enhances the quality of life and where mass access to the provision of services is gained. Between 2000 and 2022, 1.4 billion people obtained access to safely managed water in urban areas, whilst only 0.7 billion people gained access in rural areas.[12] However, although cities are where most people will live and obtain essential services, they also pose diverse challenges, including fierce competition for resources, such as water.[13,14] Cities create huge issues such as pollution, damage to the ecosystem, loss of biodiversity, and land-cover change, whilst some social aspects are also challenging in cities, such as extensive commuting times, crime, violence and social disparities.[15–17]

This chapter describes how cities may be analysed quantitatively in many ways. Firstly, by comparing between cities, it is possible to detect if there is a pattern among them. For example, if a city has more restaurants than other cities or whether it receives more tourists. Thus, the first step is to construct an urban indicator (say, the number of restaurants). Secondly, by correlating those patterns to some covariates. For example, we could analyse whether a city has more restaurants if it is a coastal city or a tourist destination, so the covariates considered could be the distance to the nearest beach or the number of tourists in a year. The most common and perhaps significant covariate to analyse is city size, so some correlations between the number of inhabitants and some urban indicators are often considered. Finally, we analyse the same indicator at more granular levels. For example, we analyse whether restaurants in the city are homogeneously distributed or if there are neighbourhoods with more restaurants. Although many spatial covariates could be considered at the city level, the distance to the city centre will often show a marked pattern and discrepancies between central parts of a city and remote locations.

2. How to analyse cities using complex systems

Our understanding of societies has rapidly evolved in recent decades, mainly due to new data that opened ways of observing patterns, often at a global scale.[2,18–20] For example, nearly 20 years ago, an experiment was conducted by tracking where banknotes appear. Tracing two successive appearances of the same bill (often appearing in distinct locations, suggesting that a person travelled from the first to the second location), relevant patterns of human travel were quantified and mapped for the first time at such large scale.[21] Similarly, data corresponding to mobile phone calls or social media has been used to model mobility patterns, opinion dynamics, and other social patterns such as segregation or polarisation.[22–25] This way of observing societies through the lens of large amounts of data is possible due to the existence of novel ways of capturing information, the increasing processing power, and new models that can be used to analyse them.

At the urban level, novel data and models have also contributed to understanding how we live in cities. For example, credit card data was used to understand lifestyles in urban populations.[26] Also, social media and real-time GIS data enabled forecasting mobility for traffic control, improving public transport systems and designing "smart" cities.[27,28] Percolation models and real-time high-resolution GPS data have improved our understanding of real traffic jams.[29,30] Similarly, the collective pedestrian movement has been analysed from the perspective of a self-organising process.[31,32] These types of analysis require large amounts of data and mathematical models based on complex systems. Thus, this way of observing cities is only possible in the past few years.

New data and models have substantially changed our way of observing the infrastructure in cities as well. One interesting dataset comes from a collaborative project called Open Street Map.[33] The project aims to create a free, editable map of the world, built by a community of mappers who contribute data about roads, railway stations, and more. The data from Open Street Map is freely available and can be used by anyone for various purposes, including navigation, research, and application development. Given its collaborative nature, it has many positive sides, particularly that users maintain it, so it creates an open-access dataset. However, it also has some negative aspects, particularly that not all roads or infrastructure have been mapped and that nobody can confirm whether the infrastructure is still there after it has been added.[34]

Besides road infrastructure, it is now possible to analyse the location and footprint of all buildings in many regions of the world.[35,36] Obtaining data corresponding to millions of buildings requires AI tools, high-resolution satellite imagery, internet capacity for sharing massive amounts of data, and computing power to process more than 1.8 billion building detections. For many limitations, this was an impossible task just a few years ago. Further, building data can be combined with other novel sources, such as precipitation, solar radiation, wind speed and other climatic variables and terrain attributes to model correlations between human settlements and weather.[35,37–39]

It is possible to analyse the structure of a city using buildings' footprint data, considering its visible infrastructure (as underground infrastructure is not detected). This type of data does not usually include roads or transport infrastructure such as railways. Instead, it contains the polygon that forms the footprint of each building (see, for example, all buildings in Basel, in Fig. 1, produced with individual building data[36]). Some interesting calculations are possible, such as the orientation of some neighbourhoods or the building size distribution. The data opens new ways to analyse cities based on what is constructed.

Building data opens new ways to observe cities and how they form a complex system. For example, when you walk through a neighbourhood, the buildings and houses often seem to be of similar height. Perhaps it is

Figure 1.   Buildings detected in Basel, Switzerland. The colour corresponds to the (log) height of buildings. Data from Ref. [36].

reasonable to think that if most of the structures of a neighbourhood are two to three stories tall, adding a much taller building would stand out and disrupt the visual flow of the area. Conversely, if a neighbourhood consists mainly of high-rise buildings, a single-story house may look out of place. Although many reasons may explain why this may happen (such as zoning regulations, architectural styles, historical development, economic factors, and more), it is relevant to ask first if this is the case. Are neighbourhoods formed mostly of buildings of similar height?

One way to answer this question is by using building data. Consider, for example, all buildings constructed in Vichy, France and their heights (Fig. 2). For building j, we express its height in metres as h_j and its age in years as g_j. Then, for building j, we can find which one is closest to it, say k, and compute the difference between their heights $d_j = |h_j - h_k|$. If we compute this metric for all buildings, then we can analyse their distribution and compute some metrics, like the average.

In Vichy, the two nearest buildings have an average difference in their heights of 2.2 m. Although there are some pairs of buildings that have a much bigger difference, we observe that most nearby buildings have a similar height. However, is 2.2 m a small difference? Could this be the result of a random distribution of building height in the city? We test this hypothesis as follows. We take all buildings in the city but permute their

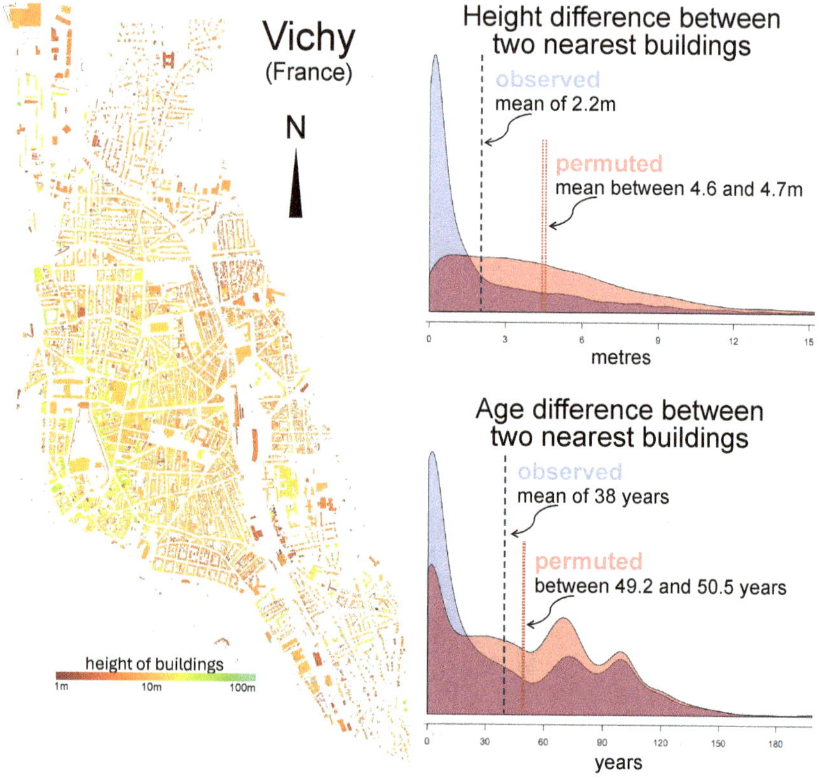

Figure 2.   Buildings in Vichy, France. Vichy is a spa town in the centre of France with roughly 25,000 inhabitants and with many centuries of history. The colour corresponds to the (log) height of buildings. Data from Ref. [36]. The right panels correspond to the distribution of the height difference between any pair of nearest buildings (top) and the age difference between them (bottom). The blue distribution corresponds to the observed display of a city, and the red distribution corresponds to a permuted city, where we re-arrange the height (or age) of all buildings in a random location.

heights randomly. That is, the city has the same layout, but buildings are assigned a height from one building in that city. Then, we measure again the difference in height between the nearest buildings, say d_j^P, and analyse their distribution and compute the average. If the distribution of d_j and the permuted city d_j^P is similar, then the observed height of buildings might be observed simply by randomness. However, this is not the case. In Vichy, the permuted city has, on average, more than double the height difference between the two nearest buildings. We can repeat the same permutation and obtain a similar difference between the nearest buildings.

A common practice to construct confidence intervals is to repeat the same type of permutation many times (usually 1,000 times is enough), drop the top and bottom extremes and keep the rest as what could be observed by randomness. By considering permutations of the height of buildings in the city, we obtain that, by chance, we should observe a difference of 4.6 to 4.7 m between the nearest buildings, more than twice the observed difference of 2.2 m (Fig. 2). Thus, at least in Vichy, France, we are sure that some neighbourhoods have tall buildings and some have shorter buildings and thus, the buildings and houses often seem to be of similar height.

We can do a similar test for the age of buildings and houses in a city. If we take the difference in age between the two nearest buildings, say $\delta_j = |g_j - g_k|$ and analyse its distribution, we can analyse if neighbourhoods have buildings of similar age. In Vichy, the average difference in age between the two nearest buildings is 38 years (Fig. 2). With a permutation of the age of buildings, we obtain that a difference between 49.2 and 50.5 years should be observed. Thus, nearby buildings also have a similar age. Notice, however, that the age between the nearest buildings is much closer to the one observed by randomness than the height. Thus, even if two nearby buildings are constructed many years apart, they tend to preserve similar heights.

Building data at the street and at the neighbourhood level has opened new ways of observing cities globally. Are buildings of similar height across neighbourhoods also observed elsewhere? Particularly in more modern cities? Or in bigger cities? Yet, for many of the metrics that can be studied at the city level, there is a fundamental issue that needs to be considered: cities vary too much in size. One of the most critical elements to consider in the analysis of cities is their population. The number of inhabitants of urban areas might vary from a few thousand people to a large metropolis of millions of people. Thus, it is relevant to understand how cities change depending on their size. For example, we could simply consider the number of buildings as an urban metric. But that number has a very close relationship with the population of that city. Cities with thousands of people have thousands of buildings, whereas cities with millions of people also have millions of buildings. Thus, it is worth constructing urban indicators that do not depend on the population in an obvious manner.

3. Urban scaling

In many ways, a large city is not just a scaled version of a small city. Most

                    R. Prieto-Curiel

components and aspects of a city vary with its size. And this variation might not scale linearly with the population. For example, take two cities, C_1 and C_2, with a population of P_1 of one million people and P_2 of two million people (so $P_2/P_1 = 2$). Then, consider its constructed volume, V_1 and V_2. Should it follow that $V_2/V_1 = 2$ as well? This scenario would be the case if every person requires a fixed amount of constructed infrastructure, say μ m^3. Then, what about the constructed surface in cities? Consider the surface S_1 and S_2. Should it follow that $S_2/S_1 = 2$ as well? If large cities are not simply a scaled version of small cities, then we can consider three scenarios (Fig. 3). One scenario is when a city only has vertical growth. In that case, we would obtain that $S_2/S_1 = 1$. A second scenario is where the growth is only horizontal, so we put a city like C_1 next to a copy of itself to make city C_2. In such case, $S_2/S_1 = 2$. However, a third scenario is when both height and surface grow at similar rates, so a large city is a scaled version of a small city. In such a case, it is possible to show that $S_2/S_1 = 2^{2/3}$.

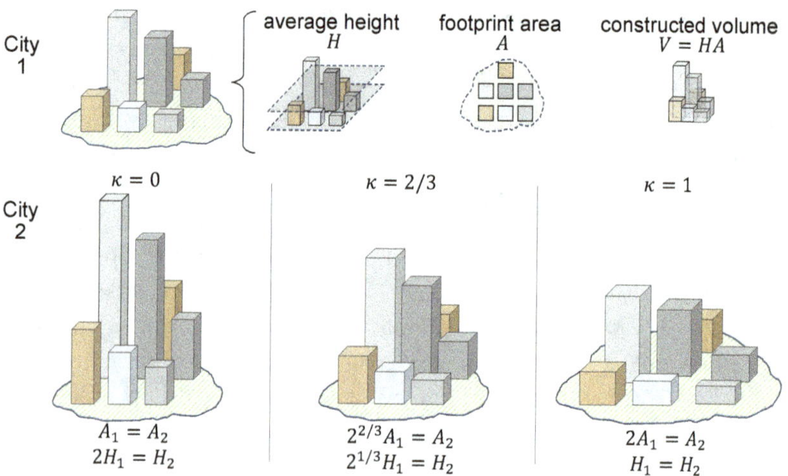

Figure 3. It is possible to capture the average height of buildings in a city, as well as the constructed surface and volume. For a city with twice the population (bottom), we may assume that it also has twice the constructed volume. Yet, there are many configurations in the way the city may expand its constructed volume, either by increasing the height, the surface or both.

The growth of the city surface as it doubles its population can be expressed by a factor of 2^κ, where $\kappa = 0$ indicates vertical growth, $\kappa = 1$

is a horizontal expansion, and $\kappa = 2/3$ corresponds to a scaled growth in all dimensions. Also, values above $\kappa = 1$ suggest not only horizontal growth but also that the city adds even more infrastructure with more population (for example, an airport, a stadium or an industrial park). In reality, most scenarios are reasonable and depend too much on the local conditions and the budget. In Hong Kong, for example, the population nearly doubled between 1970 and 2020, but in a region where the space is severely constrained, most of its growth was vertical (so $\kappa_{HK} \approx 0$). The contrary has happened in some cities in Africa, where most of its growth has been horizontal (so $\kappa_{Af} \approx 1$).[18] Analysing and comparing cities consists of constructing indicators that capture the main elements of those different configurations.

In general, urban scaling studies are interested in capturing how the varying population affects aspects such as its surface, its infrastructure or its social life. Such studies have shown, for example, that people from larger cities tend to produce more patents, have a higher income per capita, and some types of diseases, but are also prone to more congestion, road accidents, and some diseases.[40–43] Smaller cities usually require more infrastructure per person, such as the road surface or the petrol stations.[44] People from larger cities migrate less and are more likely to return after moving.[45]

3.1. *Quantifying urban scaling*

One mathematical expression that helps capture the impact of city size is the equation

$$Y_i = \alpha P_i^{\beta}, \tag{1}$$

where Y_i is the variable of interest for city i and α and β are unknown parameters. For example, Y_i could be the number of restaurants in the city i. We obtain the coefficients α and β, usually through a regression. With $\beta > 1$ the variable of interest is called "superlinear" and indicates that large cities have higher values of the variable Y per capita (since the per capita rate Y_i/P_i is given by $\alpha P_i^{\beta-1}$). With $\beta < 1$, results are "sublinear", and with $\beta \approx 1$, city size has little or no impact on the per capita rate of that city. Therefore, the coefficient of interest is usually β and values above or below $\beta = 1$ are critical. When using Eq. (1), we are not assuming a linear correlation with the population, although it can also be a result if values of β are close to one.

For some social indicators in the USA, $\beta = 1.15$ was obtained. The coefficient means that if we compare two cities, C_1 and C_2, where the population of C_2 is two times the population of C_1, the expected values of Y_2 are approximately $2^{1.15} = 2.22$ times larger. From that expression, a per capita basis is that in the city C_2, there are 11% more of Y than in city C_1. Values of $\beta \in (0,1)$ indicate that the corresponding variable Y increases with city size but at a slower rate than population (so the per capita ratio decreases with size). Finally, the same mechanism can be used to detect if some city indicator decreases with a larger population (for example, green areas or wildlife), when $\beta < 0$.

3.2. *Why do we observe urban scaling?*

Before trying to understand why we observe urban scaling, it is relevant to highlight that for many scaling properties, it remains unclear whether the coefficients observed in one country can be exported to other parts of the world or whether some of the results depend on how cities are defined or what is being measured.[46,47]

Although some aspects related to urban scaling are still unclear, some explanations of why we could observe them are interesting. Some of the urban scaling properties are usually attributed to economies of scale, based on a utility maximising mechanism for some economic agents living in those cities.[6,44,48] Also, by drawing some similarities between the body mass and the metabolic rate of biological systems to cities and their population, some scaling related to urban infrastructure has been delineated.[44] Some explanations of why we observe some scaling have been related to the resources needed to introduce a new member into the city,[44] the number of interactions between different people,[49] transportation costs, economies of scale,[50] the selective migration of highly productive people,[40] the distribution of GDP,[51] among many other theories.[52] However, it remains unclear why we observe some of the scaling properties among cities, whether they are universal and persistent, or whether they are mostly observed in urban areas in the USA.

Urban scaling and its implications are quite relevant, particularly for some of the scaling coefficients related to social life. For example, it has frequently been argued that large cities suffer more crime than small cities.[44,53,54] For instance, larger Brazilian cities report more homicides per 100,000 inhabitants than smaller ones, with similar trends observed for homicides in Colombia, burglaries in Denmark, and property crime in the

USA.[55–58] Yet, even if we accept that crime scales superlinearly with city size, would that "condemn" large cities to always be insecure? Can we not construct safe metropolitan areas and big cities? Indeed, opposite indicators have also been observed, where larger cities are less violent, such as murders and homicides in India and burglary in South Africa.[57,59] Thus, urban scaling is an interesting angle from which to observe huge disparities in city size, but results need to be taken with caution and with an understanding of its possible implications.

4. Scaling infrastructure of cities

Imagine a city with n buildings, each occupied by m people, with the same area (say, one m^2). If buildings are arranged one next to the other to minimise the average distance between them, they would be arranged circularly and compactly, occupying a surface of n m^2. For a sufficiently large number of buildings, they will form approximately a circle with an area of n m^2. The mean distance between those buildings would be $128\sqrt{n}/(45\pi)$.[18] Therefore, in a city with a circular and compact shape, distances grow with the square root of the population. With the dataset that gives the location of buildings in a city, it is possible to measure the distance between each pair of them. Thus, we can take all the buildings that correspond to some city, measure the distance between every pair of them, and then compute their average. For a set of cities, $i = 1, 2, \ldots, m$ and their buildings, we can measure the average distance between the buildings for each city, D_i. Then, we aim to understand the correlation between the population in cities, P_i and the mean distance between its buildings, D_i. With the equation

$$D_i = \alpha_D P_i^{\beta_D}, \tag{2}$$

we express the observed mean distance in cities in terms of the population of each city. There are two unknown coefficients, α_D and β_D, which capture the impact of population. For estimation, we can take the logarithm on both sides of Eq. (2) and obtain a linear expression in terms of $\log \alpha_D$ and β_D. As a baseline, we know that a coefficient of $\beta \approx 1/2$ suggests that cities grow horizontally and follow similar patterns. A coefficient $\beta < 1/2$ suggests a more round or compact growth in terms of surface (possibly with vertical expansion). With $\beta > 1/2$, cities expand mostly in horizontal ways, increasing distances faster than the natural reasons attributable to population. By considering the distance between buildings of more than

6,000 African cities, it was obtained that the coefficient $\beta_D = 0.543$, so distances in cities grow at a faster rate than the horizontal case $(1/2)$.[18]

Why are distances in cities in Africa above the threshold of $\beta_D = 1/2$? There are four explanations related to the way cities expand. Either 1) large cities have bigger buildings, 2) they grow in non-circular ways, 3) they expand in less compact manners, or 4) the actual number of buildings in cities grows faster than the population. We can decompose the surface of a city into four multiplicative components: B_i is the number of buildings in the city, A_i is their average area, S_i is the *Sprawl* (corresponding to the space between buildings, where smaller values are observed in compact cities, and E_i is the *Elongation* of a city, where smaller values indicate a round shape and higher values suggest a more elongated footprint. Thus, the area of city i is expressed as $B_i A_i S_i E_i$ and the mean distance between two buildings is given by

$$D_i = \frac{128}{45\pi} \left(\underbrace{B_i A_i}_{\text{footprint}} \; \underbrace{S_i E_i}_{\text{shape}} \right)^{1/2}. \tag{3}$$

The BASE model decomposes distances into four multiplicative components.[18] Inspired by an ellipse, we measure the Elongation of a city. The mean distance between two random points inside an ellipse has no closed solution, but an approximation can be constructed by considering the ratio between the major and the minor axis.[60] Thus, we define the Elongation E_i as

$$E_i = \frac{\sqrt{\pi} M_i}{2\sqrt{B_i A_i}}, \tag{4}$$

where M_i is the longest distance between any two buildings (so the major axis of the "ellipse" formed by the city) and where $2\sqrt{B_i A_i/\pi}$ is the smallest possible diameter of a circle with $B_i A_i$ as a footprint (so the minor axis of that "ellipse"). We obtain a coefficient $E_i \geq 1$ concerning the number or area of buildings, where $E_i = 1$ indicates a perfectly round city. Here, if a city is a scaled version of another, the elongation E_i should remain the same. For example, in a city that has four times the number of buildings, then distances also grow, including doubling the maximum distance and doubling the smallest diameter, so E_i remains the same. From the data corresponding to the buildings of a city, we can directly measure the number of buildings, B_i, their mean area A_i, the mean distance D_i and the maximum distance M_i, so we can also compute the Elongation E_i and obtain the Sprawl, given by

$$S_i = \frac{45^2 \pi^{3/2} D_i^2}{2^{13} M_i \sqrt{B_i A_i}} = \gamma \frac{D_i^2}{M_i \sqrt{B_i A_i}}, \tag{5}$$

for $\gamma = 45^2 \pi^{3/2}/2^{13} \approx 1.38$. The most relevant part is that the components of the BASE model are comparable across cities of different sizes. One way to interpret the Elongation and the Sprawl of a city is that the mean distance between buildings in cities increases proportionally to $\sqrt{S_i E_i}$, so if a city has an elongation value of $E_i = 4$ then the mean distance between buildings is double because the city is Elongated. Similarly, for the Sprawl S_i and its impact on distances.

To analyse the impact of the population, we can compute the scaling coefficient of each component of the BASE model separately. First, in terms of the footprint of a city (BA). In Africa, the number of buildings within a city grows with population, with roughly one extra building for every 2.6 people.[18] However, that number decreases slightly with size, reflecting some shared infrastructure, with $\beta_B = 0.981 \pm 0.007$. The average size of buildings also grows slightly with city size, with a coefficient $\beta_A = 0.067 \pm 0.005$. That means that if city C_2 has twice the population of the city C_1, then the average size of their buildings is $A_2/A_1 = 2^{0.067} \approx 1.05$. Thus, in a city with twice the population, the average area of buildings increases by about 5%.

In terms of the shape of the city (SE) there are also some effects in terms of city size. The Sprawl and the Elongation across cities in Africa vary considerably, from cities that are almost a compact circle to highly elongated and sprawled shapes (Fig. 4). Yet, we find only a minor effect of city size on the Elongation or the Sprawl of a city, meaning that as cities grow, they are not forming rounder or more compact shapes, on average. In Africa, we observe that $\beta_E = 0.005 \pm 0.005$, meaning that cities are equally round if they are small or big cities. Also, we observe that $\beta_S = 0.011 \pm 0.006$, meaning that cities are slightly more sprawled as they grow.

Combining the effect of the number of Buildings, their increasing Area, and the negligible effect of the Sprawl and Elongation, we can express the distance between buildings by its four components (from Eq. (3)). Thus, the fact that in Africa, the obtained scaling coefficient for the mean distances is $\beta_D = 0.532$ is mostly because the number of buildings increases sublinearly (with $\beta_B = 0.981 \pm 0.007$), but also because the size of their buildings increases (with $\beta_A = 0.067 \pm 0.005$), their Sprawl remains almost

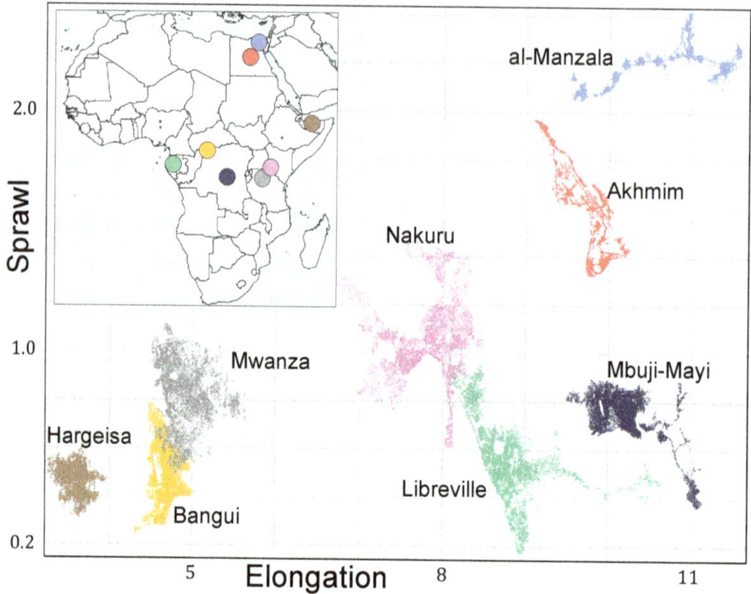

Figure 4. Eight cities of similar population in Africa are plotted using the same scale. Each building in the city is represented by a tiny dot, thus revealing the internal patterns of the city. The horizontal axis is the Elongation, and the vertical axis is the Sprawl of each city. Cities are centred on their corresponding Elongation and Sprawl. Round and compact cities are on the bottom left, whereas elongated and sprawled cities are on the top right corner.

the same (with $\beta_S = 0.011 \pm 0.006$), and their Elongation also remains almost the same (with $\beta_E = 0.005 \pm 0.005$). Combining all coefficients in Eq. (3), we get that

$$D_i \propto (B_i A_i S_i E_i)^{1/2}$$
$$\propto P_i^{\frac{0.981+0.067+0.011+0.005}{2}} = P_i^{0.532},$$

and thus, distances in large cities grow slightly faster than the square root of its population, with exponent $\beta_D = 0.532$.

With the system of Eqs. (3), (4), and (5) it is possible to imagine a city with a population of 80 million inhabitants. If present trends continue, it will have 18 million buildings (roughly two-thirds of the number of buildings there are today in Nigeria) with a footprint of 175 km^2 and an average distance between buildings above 85 km. The burden on city dwellers of such distances could force the city to grow roughly circular, compact and more vertical.

Cities face two opposing forces as they grow. People demand more infrastructure, so more and larger buildings. However, as the population increases, distances grow and become critical, so cities experience intense competition for space and make better use of it, which results in less elongated cities.[61]

4.1. *The Line in Saudi Arabia*

Saudi Arabia plans to construct a new city, home to 9 million people, called The Line. The most relevant aspect is its form, a line with a surprising length of 170 km, enough to travel from the East to the West coast of Italy.[62] It aims to host millions in a footprint of only 34 km^2. Unlike any other city, The Line is conceptually a straight line stretching from the Red Sea, 170 km East. The whole city is planned to be composed of two unbroken and joint lines of 500 m high skyscrapers and a width of 200 m. The buildings in The Line will be taller than the Empire State Building and all constructions in Europe, Africa and Latin America (Fig. 5).

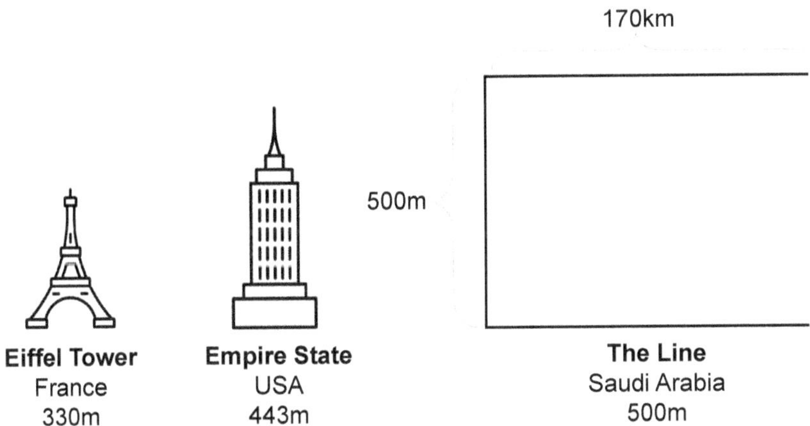

Figure 5. The Line is a planned city in Saudi Arabia. Its plan includes an elongated infrastructure of 170 km long, spanning a 500 m tall continuous building.

In The Line, we can consider the whole concept as a single building (so $B_L = 1$ building), with a surface of 34 km^2 (so $A_L = 34$ km^2), with a maximum distance of 170 km from one extreme to the other (so $M_L = 170$

km). The Elongation of The Line is

$$E_L = \frac{\sqrt{\pi} M_L}{2\sqrt{B_L A_L}} = 25.8, \tag{6}$$

which is an extremely large value of its Elongation. Further, the average distance between two random locations in a line of length l is $l/3$, so we get that $D_L = 170/3$ km. Thus, we get that the Sprawl of The Line will be

$$S_i = \frac{45^2 \pi^{3/2} D_L^2}{2^{13} M_L \sqrt{B_L A_L}} = 4.5, \tag{7}$$

also highly sprawled. Therefore, in The Line, their extreme Elongation and Sprawl contribute to the distance between people in the city being $\sqrt{S_L E_L} \approx 10.7$ times longer than on a circular city.[62]

Although the shape of The Line has some negative aspects, including its extreme elongation and the fact that it requires extremely tall and costly infrastructure to achieve such high densities, there are at least two positive aspects regarding the city. First, they aim to minimise the footprint of the city. Thus, the plans for The Line start by recognising that urban sprawl has negative impacts on the planet. But secondly, they aim to construct a car-free city. Thus, they also recognise that modern urban mobility should rely primarily on active mobility for short distances and in public transport for longer journeys.

4.2. Urban scaling and ownership

We can analyse ownership at the city level using the principles of urban scaling. Consider, for example, how many cars are there in a city. If the probability that a person owns a car does not vary according to the city size in which they live, then we should observe a fixed ratio between the number of people and the number of cars (for example, one car for every four people). Yet, if city size has an impact on the way we move, then it could also affect the number of cars we own. Considering 75 metropolitan areas in Mexico shows that for every 100 people, there are roughly 15 domestic cars (including vans). Yet, the data also reveals some scaling correlations between car ownership and city size (Fig. 6).

We estimate the scaling of the number of cars in city i by expressing

$$V_i = \alpha_V P_i^{\beta_V}, \tag{8}$$

for some values of α_V and β_V. We obtain that $\alpha_V = 0.0434 \pm 0.0259$ and that $\beta_V = 1.0972 \pm 0.0359$. Since β_V is slightly greater than one, in

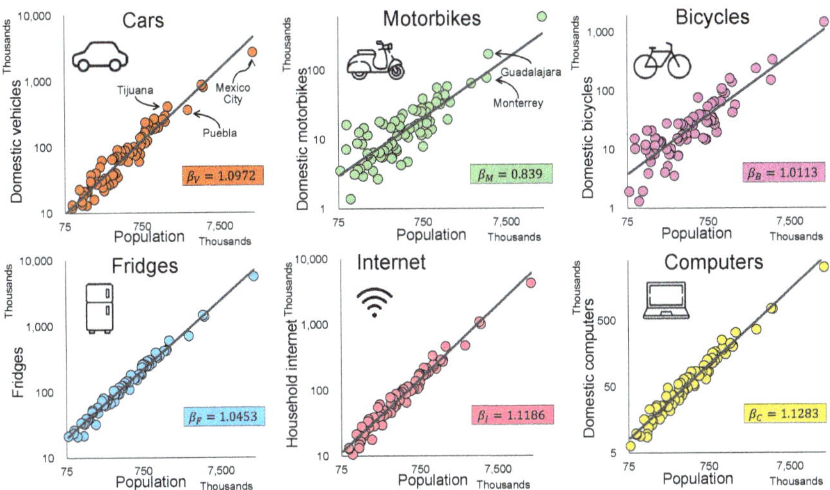

Figure 6. Number of cars and vans (vertical axis), the number of motorbikes, the number of bicycles, fridges, households with internet and computers in 75 metropolitan areas in Mexico according to the population of each city (horizontal axis). Both axes are represented on a logarithmic scale. Data from the 2020 census.[63]

large cities, there is a higher number of cars per person. If we compare two cities in Mexico, C_1 and C_2, where the population of C_2 is two times the population of C_1, the expected number of cars in C_2 is approximately $2^{1.0972} = 2.14$ times larger. Therefore, in city C_2, the probability that a person owns a car is 7% higher than in city C_1.

There is a similar pattern for the ownership of motorbikes and bicycles. In Mexico, there are 2.6 motorbikes for every 100 people, but the rate is not the same across the whole country. We find that for motorbikes, $\alpha_M = 0.2402 \pm 0.2976$ and $\beta_M = 0.8389 \pm 0.0618$, so there are fewer motorbikes per person in larger cities. Also, in Mexico, there are 5.7 bicycles for every 100 people, and we find that $\alpha_B = 0.0437 \pm 0.0627$ and $\beta_B = 1.0113 \pm 0.0683$, meaning that if we consider the number of bicycles per 100,000 inhabitants, we find a similar number in small or big cities (Fig. 6). City size influences mobility choices, the feasibility of different modes of transport and even the ownership of cars and motorbikes.[64]

It has been observed for cities outside the US, that its size has a significant influence on transport patterns.[64] In small cities outside the US, Active mobility (including walking, cycling, skating and others) and Car journeys are more common, but public transport options are limited, and for large cities, there is a shift towards public transport as the primary mode of transportation.[64]

We can also analyse the impact of city size and ownership of other home appliances. For example, in Mexico, there are roughly 27 Fridges for every 100 people, but the rate varies with city size. We observe that for the number of Fridges, $\alpha_F = 0.1589 \pm 0.0376$ and $\beta_F = 1.0452 \pm 0.0163$, meaning that in larger cities, there is slightly higher ownership of a Fridge (Fig. 6). Similarly, for households with access to the Internet, $\alpha_I = 0.0395 \pm 0.0151$ and $\beta_I = 1.1186 \pm 0.0248$, meaning more access in larger cities. Finally, for the number of domestic Computers, we find that $\alpha_C = 0.0251 \pm 0.0102$ and $\beta_C = 1.1282 \pm 0.0263$, so there is also a higher number of computers for domestic use in large cities.

Again, if we compare two cities, C_1 and C_2, where the population of C_2 is two times the population of C_1, the expected number of Fridges per person is 3% higher in C_2. Also, in the city C_2, there is a 9% higher probability of having domestic access to the Internet and a 9% higher probability of owning a Computer for domestic use (Fig. 6).

Some of the results of the urban scaling studies need to be taken with caution. For example, we observed the number of vehicles in cities of varying size (Fig. 6). We detect that since $\beta_V > 1$, then in large cities, there is a higher number of cars per person, and Fig. 6 somehow confirms this scaling behaviour. Yet, both axes are displayed on a logarithmic scale, so the "small" error that the model has for the largest city in the country (Mexico City) seems of a similar magnitude to the error produced across other cities. However, if we observe the same model without the logarithmic scale, the errors produced for that city become more evident (Fig. 7). The model produces an estimated number of cars in Mexico City of 4.6 million, which is 1.9 million above the observed number of cars. Thus, the model overestimates the number of cars by a whopping 69%. Using a logarithmic transformation reduces errors that are systematically produced by scaling models.

The same type of error occurs for the number of motorbikes and bicycles in the country. According to the scaling model, there should be 49% fewer motorbikes in Mexico City and 25% fewer bicycles (Fig. 7). Thus, it is relevant to consider that some aspects of a scaling model might be the result of some visual display, where the results may be perceived as being obvious.

If instead of analysing the raw number of cars in a city, we look at the number of cars per 100 people, the results become less evident (Fig. 8). Now, the number of domestic vehicles per 100 people does not seem to have

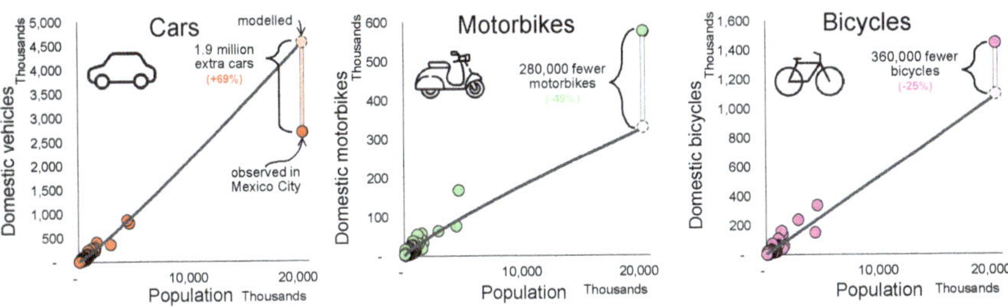

Figure 7. Number of cars and vans (vertical axis), the number of motorbikes and the number of bicycles in 75 metropolitan areas in Mexico according to the population of each city (horizontal axis).

an obvious link with city size, and the same happens with the number of motorbikes and the number of bicycles per 100 people.

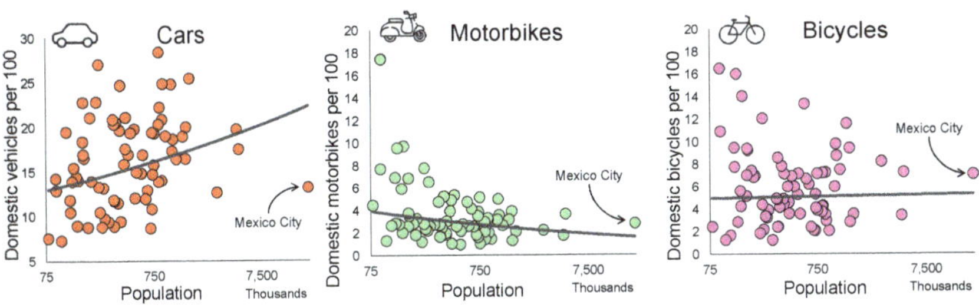

Figure 8. Number of cars and vans (vertical axis), the number of motorbikes and the number of bicycles per 100 people in 75 metropolitan areas in Mexico according to the population of each city (horizontal axis). The horizontal axis (population) is displayed on a logarithmic scale.

Finally, there is also another issue related to the observations considered. In the case of the 75 metropolitan areas of Mexico, we assign the same weight to the information provided by Mexico City (a metropolitan area with 21 million inhabitants) and to the information provided by Rioverde, a metropolitan area with less than 100,000 inhabitants. In fact, the metropolitan area of Mexico City has a larger population than the smallest 57 metropolitan areas combined. Should the results assign the same weight to all observations? One option is to consider a weighted regression, where more relevance is given to cities with a larger population, assuming that these observations provide more information. When we consider a weighted regression to estimate the scaling coefficients, we obtain that

$\beta_V^{(w)} = 0.9582 \pm 0.0178$, meaning that under this scope, the number of vehicles per city is sublinear. Similarly, the number of motorbikes is linear, with $\beta_M^{(w)} = 0.9951 \pm 0.0274$, and the number of bicycles is superlinear, with $\beta_B^{(w)} = 1.0897 \pm 0.0310$.

Estimating the scaling coefficients is challenging. Under different methods, the observed variable can be considered either superlinear (as in the case of cars in cities in Mexico) or sublinear (as in the case of cars in cities in Mexico, but when coefficients are estimated using a weighted regression). Therefore, the estimation of the scaling parameters can reveal conflicting results. However, scaling studies also have other challenges related to their approach besides the estimation of the parameters. Similar issues have been observed elsewhere, where some scaling even change the sublinearity to superlinearity.[65]

4.3. *The challenges of urban scaling studies*

Urban scaling studies have many challenges. Firstly, defining urban agglomerations is difficult and depends on considerations and parameters.[46,66] What is actually a city and how to compare between them is not at all a trivial aspect. Urban scaling values are not universal, and they significantly depend on the way one constructs cities and the variables considered.[65] The way cities or metropolitan areas are defined may alter results.[67–69] This issue is particularly relevant when cities from different countries are compared. Particularly in some developing regions, census data might be outdated and have critical implications. For example, the latest available census in Somalia was carried out decades ago, but in Kenya, a neighbouring country, the most recent census was in 2019. Thus, comparing cities in East Africa may be challenging.

One of the biggest challenges of urban scaling studies is that many of the results obtained so far correspond to observations from the USA. However, more data from other regions of the world has opened new studies that challenge the universality of scaling values. For example, observing the shape of 5,000 cities in Africa revealed that their surface is correlated linearly with city size.[18] However, a non-trivial growth of the average size of buildings, with an approximately linear increase in the number of buildings and urban form indicators, increases the expected value of the distance between different parts of the city.[18]

Yet, new technology is also opening ways to observe the population and cities. For example, satellite images are now a valuable tool for observing population settlements and estimating the number of inhabitants at a very refined scale. Thus, this type of technology is often providing more accurate information than a census that was carried out decades ago. Using this type of input, a novel dataset that delineated all African cities is now available. Africapolis maps the location and boundaries of more than 5,000 urban agglomerations based on the same definition for an urban agglomeration at a continental level.[70] It is based on the infrastructure of a city, and based on aerial images, it delineates a city if its buildings are less than 200 m apart. Thus, scaling studies outside the USA are now more feasible.

Also, beyond numerous urban scaling correlations, it is crucial to understand the driving forces and mechanisms behind them. For example, it is possible to measure greenhouse gas emissions at the city level and detect that large cities produce fewer emissions per person than smaller cities.[71] Yet, although the result is relevant by itself, it is unclear why there is a reduction in larger cities. Do people in large cities make fewer trips on a regular week than people in small cities? Are there fewer emissions because people are more likely to use Public Transport? Are people less likely to drive in big cities, or is it related to infrastructure, such as the increase in electric cars in big cities? Or are bigger cities producing fewer greenhouse gas emissions because they outsource most of their industry from other places? Without an understanding of the mechanisms that cause the observed correlations, little is known with respect to their implications or what policies should be implemented to improve urban systems. If large cities produce fewer emissions only because they outsource industrial activities somewhere else, the repercussions are critically different than a reduction related to mobility.[64]

There are many ways to understand cities beyond correlational studies. For example, cellular automata and agent-based models combined with machine learning techniques have become powerful tools to simulate complex processes happening within urban dynamics.[72,73] How will a city expand due to population growth enables foreseeing the provision of services such as electricity and water infrastructure.[74] These type of models usually consider the complex interactions between agents (say, urbanites), their neighbourhood and its infrastructure. Although agent-based models have relevant challenges related to calibration and validation, they enable considering the main mechanisms and driving forces within the urban dynamics.[75]

5. Urban studies beyond scaling

Urban scaling studies tend to analyse infrastructure or social aspects of a city in terms of a single indicator, for example, in terms of income, accessibility, road insecurity or crime. Yet, for most urban indicators, there is a vast heterogeneity within the same city. For example, in terms of proximity to public transport, central areas of a city tend to be better connected and enjoy more provisions than remote areas. There is also an issue with the spatial arrangement of areas in the city. In theory, even if a metropolitan area is large, people could still live close to their destinations. However, a neighbourhood is not a small-scale version of a city. Consider, for example, the number of crimes in a city. Take, for example, the crimes reported to the Police in Mexico City (Fig. 9). Most neighbourhoods have a relatively low level of crime, and only a few locations in the city tend to concentrate most crimes.

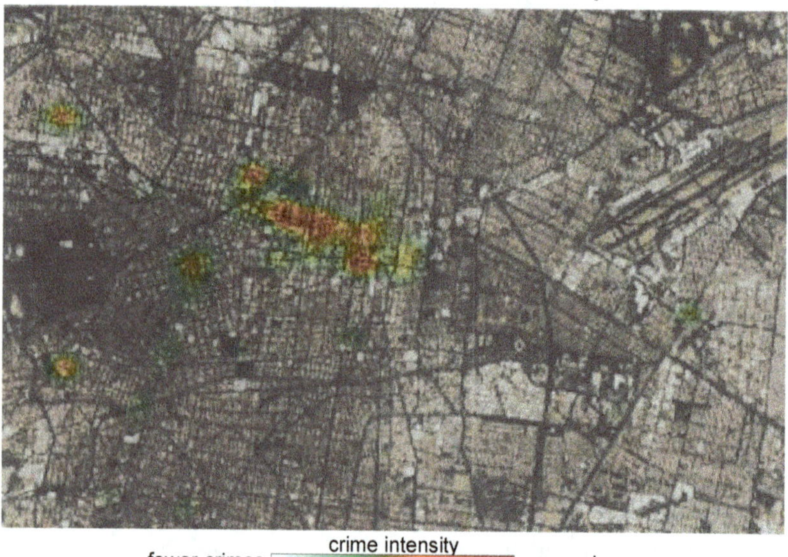

Figure 9. Intensity of crimes reported to the Police in Mexico City between 2016 and 2020. It includes only property crimes (robbery of a person, car theft, shoplifting and burglary). The intensity is computed by a 2-D kernel density based on the location of reported crimes. Neighbourhoods with fewer crimes have a transparent background, but areas with more crimes are shadowed in green, orange, and red, in that order.

Crime is highly concentrated among different neighbourhoods or street segments, resulting in hotspots of crime.[76,77] This pattern is roughly stable in time (so if we compare the hotspots from 2016, they are roughly the same as the ones in 2019) and tends to reveal areas with the highest social activities, like work or schools, transport stations and more. Thus, we see that crime is highly heterogeneous at the city level. In fact, similar to what we observe at the neighbourhood level, we see it at the personal level. It was observed that roughly 60% of the population in Mexico City is "immune" to suffering crime, meaning that they will rarely suffer any crime.[78] However, in Mexico City, some people (perhaps less than 5% of the population) are chronically victimised, meaning that they tend to suffer crime frequently.[78] Therefore, perhaps the number of crimes per person is not a relevant indicator if some people within the same city are frequently victimised whilst most people are rarely the victims of crime.

Analysing the correlation between city size and the number of crimes is relevant, so we can understand whether people in large cities tend to suffer a higher or lower crime rate than people in smaller cities. However, analysing the spatial distribution at different spatial scales within a city is also a crucial part of understanding urban dynamics, for example, by comparing different neighbourhoods in a city, street segments, or at the individual level.

Although there are many challenges to analysing cities at smaller spatial units (including the delineation of a neighbourhood and how arbitrary or political boundaries may strongly affect the results), a relevant aspect to consider is the extent to which a variable is correlated with itself through space. For example, a neighbourhood with a large number of crimes is likely close to other neighbourhoods that also have many crimes. Furthermore, the position of a neighbourhood within a city might also be a critical aspect. Some parts of a city tend to attract workers or students from other areas, whilst other parts, mainly the remote suburbs of a city, may lose large parts of their population due to daily routines.[79] In most cities, a small fraction of the neighbourhoods tend to monopolise most workers, students, shoppers and visitors.

5.1. *Comparing at a smaller scale between cities*

One way to preserve parts of the geography of cities is by looking at their position with respect to their centre. For a city, we first identify its centre. This could be, for example, the location of the main square, temple or transport station. Although the precise location of the centre could be ambiguous (for example, if there are many candidates as a central square or transport station), we pick one of those locations that may be used to determine the relative proximity of other places in the city. Based on the location of the city's centre, we measure the distance to the city centre. Let $D_{i,j}$ be the distance in km for location j in city i to its centre. With this metric, remote locations are identified by having a large $D_{i,j}$. Although the city centre could be defined in terms of more than one location, the impact on the distance to the city centre is usually small. In Paris, France, for example, the distance from Stade de France (a stadium in the North part of the city) is 7.3 km away from Louvre, 7.8 km away from Notre-Dame and 7.2 km away from Place de la Concorde (three candidates to be considered the city centre).

After we identify the centre of a city, we can quantify an urban indicator at the neighbourhood level, for example, the number of car crashes, the number of crimes, or some economic indicator, such as the number of households with access to water, with a car or a computer. Then, groups of different locations that share their distance from the centre can be formed. This way, we analyse concentric rings covering the whole city. The idea behind this is that central locations in a city have some differences from the intermediate or remote locations. For example, consider the percentage of households that own a computer. Census data from Mexico in 2020 enables us to observe the ownership of certain items at the census tract level (Fig. 10). Although there is a huge variability in terms of the number of households with a computer, it is possible to notice a pattern across all cities. In central areas, owning a computer is more frequent (with between 40 and 80% of the households with a computer). Yet, for remote locations, owning a computer is much less frequent. If the distance to the centre is above a certain threshold, then owning a computer is much less frequent (between 20 and 40% if the distance is above 30 km).

The distance to a city centre, however, is relative to the population of a city. For example, in Mexico City (a metropolitan area with 22 million inhabitants), a location 2.5 km away from the city centre is still considered a central location. However, in Puebla (a city with 3.2 million inhabitants), a location within the same distance is already distant. And in even smaller cities, 2.5 km away from the city centre would already be considered outside of its boundary. One way to remove the effect of size is as follows. Imagine a city with a population P_i, where each person occupies the same area, say $\nu > 0$ m^2. Then, its area can be expressed as νP_i. If that city is

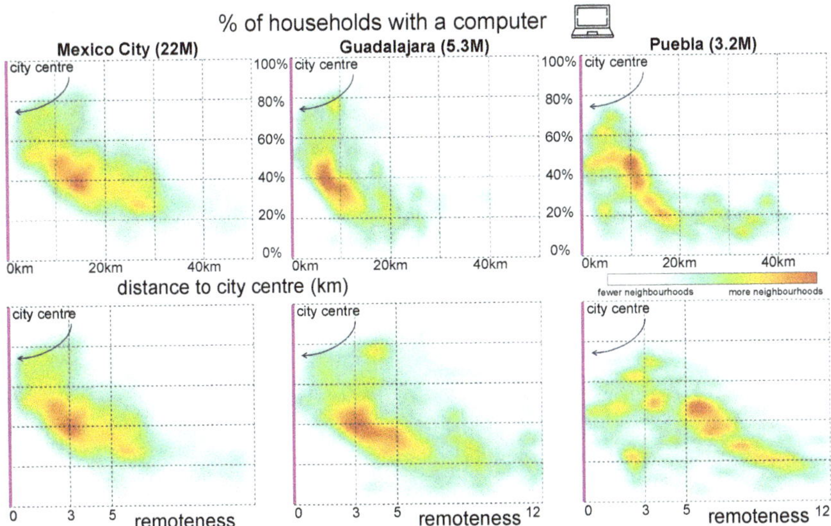

Figure 10. Taking census tracts in three cities in Mexico, we measure the number of households that own a computer (vertical axis). The centre is located at the left part of each figure, and the distance (horizontal axis) indicates all census tracts that are up to 50 km away from the centre. The left figures correspond to the metropolitan area of Mexico City (with 22 million inhabitants). The figures in the centre correspond to Guadalajara (with 5.3 million people), and the right figures are for Puebla (with 3.2 million people). The top panels show the distance to the centre, and the bottom panels show the remoteness.

arranged circularly and compactly, then we can compute the average distance between two people. If we take two random points inside that circle, the expected distance between them is

$$\frac{128}{45\pi}\sqrt{\nu P_i} = \delta\sqrt{P_i}, \tag{9}$$

for $\delta = 128\sqrt{\nu}/45\pi$, meaning that distances in a city grow with the square root of its population.[18] Thus, by considering the location j in city i and dividing by the square root of its population, we essentially remove the impact of city size.[80] We construct the *remoteness* of location j in city i, expressed as $R_{i,j}$ by

$$R_{i,j} = \frac{1000 D_{i,j}}{\sqrt{P_i}}. \tag{10}$$

The expression 10 is multiplied by 1000 to obtain numbers that are easier to manipulate. With this formula, a location in Mexico City that is 2.5 km away from its centre has $R_{X,j} = 0.5$, whereas a location in Puebla 2.5

km away from its centre has $R_{P,j} = 1.4$, so the same distance would be considered much more remote. In a small city S with only $P_S = 100{,}000$ inhabitants, 2.5 km away from its centre has $R_{S,j} = 7.9$.

The concept of remoteness provides a way of understanding cities, revealing how urban amenities tend to concentrate in central areas, leaving remote regions with fewer provisions and higher levels of social exclusion. Remote areas generally experience higher levels of poverty and social exclusion. For example, looking at the percentage of households that own a computer, there are many more similarities across cities (Fig. 10). For a small level of remoteness (for example, with $R_{i,j} < 5$), there is a wide heterogeneity in terms of owning a computer. Many regions in the central areas of a city are wealthier, and people tend to own a computer. However, in remote areas (for example, with $R_{i,j} > 5$), neighbourhoods tend to be poorer, so there are fewer households with a computer.

6. Conclusions

While urban scaling theories offer insights into how urban metrics change with population size, they are not universal laws for all cities. Rather, they provide a framework to analyse cities with varying sizes. Results provided by urban scaling studies need to be carefully analysed since they depend on many aspects including the definition of cities and parameters used. Further, if in some region it has been observed that large cities tend to have some attributes in common, it might not be a permanent pattern and there is no reason to assume it will apply to cities in other regions.

Scaling studies overlook significant heterogeneity within cities. Factors such as spatial arrangement, proximity to amenities, and socio-economic disparities play crucial roles in shaping different neighbourhoods within the same city. Analysing cities at finer spatial scales, such as neighbourhood levels, reveals essential patterns of variation. We might find clusters of similar characteristics, emphasising the interconnectedness of neighbouring areas. Moreover, considering the distance from a neighbourhood to the city centre helps capture gradients of infrastructure and socio-economic indicators, highlighting disparities between central and remote locations. This situation underscores the importance of analysing urban dynamics beyond scaling frameworks, considering spatial heterogeneity, and exploring local variations within cities. Such approaches are essential for developing targeted policies and interventions to address the diverse needs and challenges faced by different neighbourhoods within urban areas.

References

1. X. Bai, A. Surveyer, T. Elmqvist, F. W. Gatzweiler, B. Güneralp, S. Parnell, A.-H. Prieur-Richard, P. Shrivastava, J. G. Siri, M. Stafford-Smith, et al., Defining and advancing a systems approach for sustainable cities, *Current Opinion in Environmental Sustainability.* **23**, 69–78 (2016).
2. L. S. Tusting, D. Bisanzio, G. Alabaster, E. Cameron, R. Cibulskis, M. Davies, S. Flaxman, H. S. Gibson, J. Knudsen, C. Mbogo, F. O. Okumu, L. von Seidlein, D. J. Weiss, S. W. Lindsay, P. W. Gething, and S. Bhatt, Mapping changes in housing in Sub-Saharan Africa from 2000 to 2015, *Nature* (2019). ISSN 1476-4687. doi: 10.1038/s41586-019-1050-5. URL `https://doi.org/10.1038/s41586-019-1050-5`.
3. P. D. United Nations. World urbanization prospects (2023).
4. S. Angel, S. A. Franco, Y. Liu, and A. M. Blei, The shape compactness of urban footprints, *Progress in Planning.* **139**, 100429 (2020).
5. K. C. Seto, B. Güneralp, and L. R. Hutyra, Global forecasts of urban expansion to 2030 and direct impacts on biodiversity and carbon pools, *Proceedings of the National Academy of Sciences.* **109**(40), 16083–16088 (2012).
6. D. Pumain, Scaling laws and urban systems, *Santa Fe Institute, Working Paper.* **04**(02), 002 (2004).
7. D. Hoornweg and K. Pope, Population predictions for the world's largest cities in the 21st century, *Environment and Urbanization.* **29**(1), 195–216 (2017). doi: 10.1177/0956247816663557. URL `https://doi.org/10.1177/0956247816663557`.
8. D. Leon, Cities, urbanization and health, *International Journal of Epidemiology.* **37**(1), 4–8 (01, 2008). ISSN 0300-5771. doi: 10.1093/ije/dym271. URL `https://doi.org/10.1093/ije/dym271`.
9. P. Combes, G. Duranton, L. Gobillon, D. Puga, and S. Roux, The productivity advantages of large cities: Distinguishing agglomeration from firm selection, *Econometrica.* **80**(6), 2543–2594 (2012). doi: 10.3982/ECTA8442. URL `https://onlinelibrary.wiley.com/doi/abs/10.3982/ECTA8442`.
10. E. Glaeser and D. Maré, Cities and skills, *Journal of Labor Economics.* **19**(2), 316–342 (2001). ISSN 0734306X, 15375307. URL `http://www.jstor.org/stable/10.1086/319563`.
11. J. Winters, Why are smart cities growing? Who moves and who stays, *Journal of Regional Science.* **51**(2), 253–270 (2011). doi: 10.1111/j.1467-9787.2010.00693.x. URL `https://onlinelibrary.wiley.com/doi/abs/10.1111/j.1467-9787.2010.00693.x`.
12. W. H. Organization and UNICEF. Joint monitoring programme for water supply, sanitation and hygiene (JMP) (2023).
13. C. Wight, D. Garrick, and T. Iseman, Mapping incentives for sustainable water use: global potential, local pathways, *Environmental Research Communications.* **3**(4), 041002 (2021).
14. M. Flörke, C. Schneider, and R. I. McDonald, Water competition between cities and agriculture driven by climate change and urban growth, *Nature Sustainability.* **1**(1), 51–58 (2018).

15. K. Seto, B. Güneralp, and L. Hutyra, Global forecasts of urban expansion to 2030 and direct impacts on biodiversity and carbon pools, *Proceedings of the National Academy of Sciences.* **109**(40), 16083–16088 (2012). ISSN 0027-8424. doi: 10.1073/pnas.1211658109. URL https://www.pnas.org/content/109/40/16083.

16. K. Seto, A. Reenberg, C. Boone, M. Fragkias, D. Haase, T. Langanke, P. Marcotullio, D. Munroe, B. Olah, and D. Simon, Urban land teleconnections and sustainability, *Proceedings of the National Academy of Sciences.* **109**(20), 7687–7692 (2012). ISSN 0027-8424. doi: 10.1073/pnas.1117622109. URL https://www.pnas.org/content/109/20/7687.

17. P. Gong, S. Liang, E. Carlton, Q. Jiang, J. Wu, L. Wang, and J. Remais, Urbanisation and health in China, *The Lancet.* **379**(9818), 843–852 (2012). ISSN 0140-6736. doi: https://doi.org/10.1016/S0140-6736(11)61878-3. URL http://www.sciencedirect.com/science/article/pii/S0140673611618783.

18. R. Prieto-Curiel, J. E. Patino, and B. Anderson, Scaling of the morphology of African cities, *Proceedings of the National Academy of Sciences.* **120**(9), e2214254120 (2023).

19. J. Candipan, N. E. Phillips, R. J. Sampson, and M. Small, From residence to movement: The nature of racial segregation in everyday urban mobility, *Urban Studies.* p. 0042098020978965 (2021).

20. C. Guan, J. Song, M. Keith, Y. Akiyama, R. Shibasaki, and T. Sato, Delineating urban park catchment areas using mobile phone data: A case study of Tokyo, *Computers, Environment and Urban Systems.* **81**, 101474 (2020).

21. D. Brockmann, L. Hufnagel, and T. Geisel, The scaling laws of human travel, *Nature.* **439**(7075), 462–465 (2006).

22. M. C. Gonzalez, C. A. Hidalgo, and A.-L. Barabasi, Understanding individual human mobility patterns, *Nature.* **453**(7196), 779 (2008).

23. E. Bakshy, S. Messing, and L. A. Adamic, Exposure to ideologically diverse news and opinion on Facebook, *Science.* **348**(6239), 1130–1132 (2015).

24. M. Levy and J. Goldenberg, The gravitational law of social interaction, *Physica A: Statistical Mechanics and its Applications.* **393**, 418–426 (2014).

25. R. Lambiotte, V. D. Blondel, C. De Kerchove, E. Huens, C. Prieur, Z. Smoreda, and P. Van Dooren, Geographical dispersal of mobile communication networks, *Physica A: Statistical Mechanics and its Applications.* **387** (21), 5317–5325 (2008).

26. R. Di Clemente, M. Luengo-Oroz, M. Travizano, S. Xu, B. Vaitla, and M. C. González, Sequences of purchases in credit card data reveal lifestyles in urban populations, *Nature Communications.* **9** (2018).

27. W. Li, M. Batty, and M. F. Goodchild, Real-time GIS for smart cities, *International Journal of Geographical Information Science.* **0**(0), 1–14 (2019). doi: 10.1080/13658816.2019.1673397. URL https://doi.org/10.1080/13658816.2019.1673397.

28. D. Doran, Social media enabled human sensing for smart cities, *AI Communications.* **29**, 57 (2016).

29. G. Zeng, D. Li, S. Guo, L. Gao, Z. Gao, H. E. Stanley, and S. Havlin, Switch

between critical percolation modes in city traffic dynamics, *Proceedings of the National Academy of Sciences.* **116**(1), 23–28 (2019).

30. L. Zhang, G. Zeng, D. Li, H.-J. Huang, H. E. Stanley, and S. Havlin, Scale-free resilience of real traffic jams, *Proceedings of the National Academy of Sciences.* **116**(18), 8673–8678 (2019).

31. D. Helbing and P. Molnar, Social force model for pedestrian dynamics, *Physical Review E.* **51**(5), 4282 (1995).

32. D. Helbing, P. Molnár, I. J. Farkas, and K. Bolay, Self-organizing pedestrian movement, *Environment and Planning B: Planning and Design.* **28**(3), 361–383 (2001). doi: 10.1068/b2697.

33. OpenStreetMap contributors. Planet dump retrieved from https://planet.osm.org. `https://www.openstreetmap.org` (2021).

34. R. Prieto-Curiel, I. Heo, A. Schumann, and P. Heinrigs, Constructing a simplified interurban road network based on crowdsourced geodata, *MethodsX.* **9**, 101845 (2022).

35. W. Sirko, S. Kashubin, M. Ritter, A. Annkah, Y. S. E. Bouchareb, Y. Dauphin, D. Keysers, M. Neumann, M. Cisse, and J. Quinn, Continental-scale building detection from high resolution satellite imagery, *arXiv preprint arXiv:2107.12283.* **1** (2021).

36. N. Milojevic-Dupont, F. Wagner, F. Nachtigall, J. Hu, G. B. Brüser, M. Zumwald, F. Biljecki, N. Heeren, L. H. Kaack, P.-P. Pichler, and F. Creutzig, EUBUCCO v0. 1: European building stock characteristics in a common and open database for 200+ million individual buildings, *Scientific Data.* **10**(1), 147 (2023).

37. S. E. Fick and R. J. Hijmans, Worldclim 2: new 1-km spatial resolution climate surfaces for global land areas, *International Journal of Climatology.* **37**(12), 4302–4315 (2017).

38. T. Esch, E. Brzoska, S. Dech, B. Leutner, D. Palacios-Lopez, A. Metz-Marconcini, M. Marconcini, A. Roth, and J. Zeidler, World Settlement Footprint 3D-a first three-dimensional survey of the global building stock, *Remote Sensing of Environment.* **270**, 112877 (2022).

39. NASA JPL. NASADEM Merged DEM Global 1 arc second V001 [Data set]. URL `https://doi.org/10.5067/MEaSUREs/NASADEM/NASADEM_HGT.001` (2020).

40. M. Keuschnigg, S. Mutgan, and P. Hedström, Urban scaling and the regional divide, *Science Advances.* **5**(1), eaav0042 (2019).

41. C. Cabrera-Arnau, R. Prieto-Curiel, and S. R. Bishop, Uncovering the behaviour of road accidents in urban areas, *Royal Society Open Science.* **7**(4), 191739 (2020).

42. R. Louf and M. Barthelemy, How congestion shapes cities: from mobility patterns to scaling, *Scientific Reports.* **4**(1), 1–9 (2014).

43. B. Zhou, S. Thies, R. Gudipudi, M. K. Lüdeke, J. P. Kropp, and D. Rybski, A Gini approach to spatial CO_2 emissions, *Plos One.* **15**(11), e0242479 (2020).

44. L. M. Bettencourt, J. Lobo, D. Helbing, C. Kuhnert, and G. B. West, Growth, innovation, scaling, and the pace of life in cities, *Proceedings of the National Academy of Sciences.* **104**(17), 7301–7306 (2007).

45. R. Prieto-Curiel, L. Pappalardo, L. Gabrielli, and S. R. Bishop, Gravity and scaling laws of city to city migration, *PloS one.* **13**(7), e0199892 (2018).

46. E. Arcaute, E. Hatna, P. Ferguson, H. Youn, A. Johansson, and M. Batty, Constructing cities, deconstructing scaling laws, *Journal of the Royal Society Interface.* **12**(102), 20140745 (2015).

47. R. Louf and M. Barthelemy, Scaling: lost in the smog, *Environment and Planning B: Planning and Design.* **41**(5), 767–769 (2014).

48. D. Pumain, Settlement systems in the evolution, *Geografiska Annaler. Series B, Human Geography.* **82**(2), 73–87 (2000). ISSN 04353684, 14680467. URL `http://www.jstor.org/stable/491066`.

49. L. M. Bettencourt, The origins of scaling in cities, *Science.* **340**(6139), 1438–1441 (2013).

50. L. M. Bettencourt, V. C. Yang, J. Lobo, C. P. Kempes, D. Rybski, and M. J. Hamilton, The interpretation of urban scaling analysis in time, *Journal of the Royal Society Interface.* **17**(163), 20190846 (2020).

51. H. V. Ribeiro, M. Oehlers, A. I. Moreno-Monroy, J. P. Kropp, and D. Rybski, Association between population distribution and urban GDP scaling, *PLoS One.* **16**(1), e0245771 (2021).

52. F. L. Ribeiro and D. Rybski, Mathematical models to explain the origin of urban scaling laws, *Physics Reports.* **1012**, 1–39 (2023). ISSN 0370-1573. doi: https://doi.org/10.1016/j.physrep.2023.02.002. URL `https://www.sciencedirect.com/science/article/pii/S0370157323000650`.

53. E. L. Glaeser and B. Sacerdote. Why is there more crime in cities? Technical report, National Bureau of Economic Research (1996).

54. J. B. Cullen and S. D. Levitt, Crime, urban flight, and the consequences for cities, *Review of Economics and Statistics.* **81**(2), 159–169 (1999).

55. L. G. Alves, H. V. Ribeiro, and R. S. Mendes, Scaling laws in the dynamics of crime growth rate, *Physica A: Statistical Mechanics and its Applications.* **392**(11), 2672–2679 (2013).

56. A. Gomez-Lievano, H. Youn, and L. M. Bettencourt, The statistics of urban scaling and their connection to Zipf's law, *PloS one.* **7**(7), e40393 (2012).

57. M. Oliveira, More crime in cities? on the scaling laws of crime and the inadequacy of per capita rankings—a cross-country study, *Crime Science.* **10**(1), 27 (2021).

58. Y. S. Chang, H. E. Kim, and S. Jeon, Do larger cities experience lower crime rates? a scaling analysis of 758 cities in the US, *Sustainability.* **11**(11), 3111 (2019).

59. A. Sahasranaman and L. M. Bettencourt, Urban geography and scaling of contemporary Indian cities, *Journal of the Royal Society Interface.* **16**(152), 20180758 (2019).

60. M. Parry and E. Fischbach, Probability distribution of distance in a uniform ellipsoid: Theory and applications to physics, *Journal of Mathematical Physics.* **41**(4), 2417–2433 (2000).

61. M. Batty, The size, scale, and shape of cities, *Science.* **319**(5864), 769–771 (2008).

62. R. Prieto-Curiel and D. Kondor, Arguments for building The Circle and not

The Line in Saudi Arabia, *npj Urban Sustainability.* **3**(1), 35 (2023).

63. O. Website from INEGI. 2020 census data. `http://www.inegi.org.mx/` (2021). Accessed on July 2021.

64. R. Prieto-Curiel and J. P. Ospina, The ABC of mobility, *Environment International.* **185**, 108541 (2024). ISSN 0160-4120. doi: https://doi.org/10.1016/j.envint.2024.108541. URL `https://www.sciencedirect.com/science/article/pii/S0160412024001272`.

65. C. Cottineau, E. Hatna, E. Arcaute, and M. Batty, Diverse cities or the systematic paradox of urban scaling laws, *Computers, Environment and Urban Systems.* **63**, 80–94 (2017).

66. C. Rozenblat, Extending the concept of city for delineating large urban regions (LUR) for the cities of the world, *Cybergeo: European Journal of Geography* (2020).

67. G. Montero, C. Tannier and I. Thomas, Delineation of cities based on scaling properties of urban patterns: A comparison of three methods, *International Journal of Geographical Information Science,* **35**(5), 919-47 (2021).

68. E. A. Oliveira, J. S. Andrade Jr, and H. A. Makse, Large cities are less green, *Scientific Reports.* **4**, 4235 (2014).

69. H. D. Rozenfeld, D. Rybski, J. S. Andrade, M. Batty, H. E. Stanley, and H. A. Makse, Laws of population growth, *Proceedings of the National Academy of Sciences.* **105**(48), 18702–18707 (2008).

70. F. Moriconi-Ebrard, D. Harre, and P. Heinrigs. Urbanisation dynamics in West Africa 1950 - 2010. URL `/content/book/9789264252233-en` (2015).

71. M. Lu, C. Zhou, C. Wang, R. B. Jackson, and C. P. Kempes, Worldwide scaling of waste generation in urban systems, *Nature Cities.* pp. 1–10 (2024).

72. M. Batty, *Cities and complexity: understanding cities with cellular automata, agent-based models, and fractals.* The MIT press (2007).

73. J. C. Duque, N. Lozano-Gracia, J. E. Patino, P. Restrepo, and W. A. Velasquez, Spatiotemporal dynamics of urban growth in Latin American cities: An analysis using nighttime light imagery, *Landscape and Urban Planning.* **191**, 103640 (2019).

74. J. C. Duque, N. Lozano Gracia, J. Patino, and P. Restrepo, Urban form and productivity: What is the shape of Latin American cities?, *World Bank Policy Research Working Paper.* **1**(8697) (2019).

75. A. Heppenstall, A. Crooks, N. Malleson, E. Manley, J. Ge, and M. Batty, Future developments in geographical agent-based models: Challenges and opportunities, *Geographical Analysis* (2020).

76. D. Weisburd, The law of crime concentration and the criminology of place, *Criminology.* **53**(2), 133–157 (2015). ISSN 1745-9125. doi: 10.1111/1745-9125.12070. URL `http://dx.doi.org/10.1111/1745-9125.12070`.

77. Y. Lee, J. E. Eck, S. O, and N. N. Martinez, How concentrated is crime at places? a systematic review from 1970 to 2015, *Crime Science.* **6**(1), 6 (2017).

78. R. Prieto-Curiel, S. Collignon Delmar, and S. R. Bishop, Measuring the distribution of crime and its concentration, *Journal of Quantitative Criminology.* pp. 1–29 (2017).

79. M. Felson and R. Boivin, Daily crime flows within a city, *Crime Science.* **4** (1), 31 (2015).
80. R. Prieto-Curiel and C. Borja-Vega, The influence of urban morphology on water scarcity, *arXiv preprint arXiv:2402.06676* (2024).

Index

absorbing state, 83, 102
aging, 86, 87
algorithm
  $M(RT)^2$, 126
  Swendsen–Wang cluster, 25

backpropagation, 120
basic reproduction number, 98
Boltzmann machine, 122, 126
  restricted, 122, 130
buildings footprint, 184

carrying capacity, 71
civil liberties, 171, 173
coexistence, 75–77
complex system, 108, 183
  features, 164, 167
contrastive divergence, 131
correlation, 4, 72, 168, 183
crime, 202
critical
  amplitude, 12, 33, 38
  dynamics, 84
  exponent, 12, 86, 88
    hatted, 15
    starred, 42
  point, 4, 135

dangerous irrelevant variable, 19

democracy, 165, 170, 174
dimension
  anomalous, 13, 48
  upper critical, 53
disorder, 167, 170
Doi–Peliti field theory, 74, 101
dynamical system, 124
dynamics
  critical, 84
  evolutionary, 93, 94
  Glauber, 127

election quality, 171, 173
elongation, 192
epidemic, 96–99
exponent
  coppa, 8, 10, 18
  coppa hat, 17
  critical, 12
  logarithmic correction, 13
  rounding, 27
  shift, 27, 28
  starred, 42
extinction, 76, 77

feedback, 167, 170
fluctuations, 7, 73, 74, 173
Fourier modes, 43
function

activation, 123
  sigmoid, 126
  utility, 176

gradient descent, 129

Hebbian learning, 119, 125
Hessian, 135, 137
  -vector product, 145
  direction, 145
homogeneity, 10
Hutchinson's method, 139, 144

inference
  Bayesian, 121
  fiducial, 121
interface, 103–105, 107

Janssen–De Dominicis
  functional, 83, 97

Kullback–Leibler divergence, 128

linear stability matrix, 76, 77,
  100, 102
loss
  landscape, 134, 147
  projection, 136

maximum entropy principle, 121
McCulloch–Pitts neuron, 119,
  123
model
  BASE, 192
  Gaussian, 35
  Ising, 118, 122, 169
  Lotka–Volterra predator-prey,
    75, 76
  May–Leonard, 102, 106

percolation, 183
rock-paper-scissors, 100, 101
scaling, 198
SIR, 97, 99
SIS, 97
Morse lemma, 135

network
  complex, 96
  Hopfield, 122, 123

Palestine, 4
perceptron, 119
  multilayer, 120
percolation
  directed, 85, 88, 97
  dynamic isotropic, 99
population oscillation, 78, 81
principal curvature, 135, 137
pseudocritical point, 9, 27

random projection, 136, 138
randomness, 174, 176
rate equation, 72, 76
relation
  hyperscaling, 7, 58
  scaling, 12
remoteness, 205

saddle point, 138
scaling
  finite-size, 8
    corrections to, 44
  urban, 189
sprawl, 193
stability, 165, 168

theory
  game, 100, 177

Ginzburg-Landau, 31
mean-field, 31

Ukraine, 4, 108

zeros
Fisher, 20, 25
Lee-Yang, 20, 21
of the partition function, 20